NAVAJO COUNTRY

NAVAJO

DONALD L. BAARS

COUNTRY

A GEOLOGY AND
NATURAL HISTORY OF THE
FOUR CORNERS REGION

UNIVERSITY OF NEW MEXICO PRESS
Albuquerque

To Sherman A. Wengerd,
the grand old man of Four Corners geology

© 1995 by the University of New Mexico Press
All rights reserved.
First Edition

ISBN 0-8263-1587-9

Designed by Linda Mae Tratechaud

Library of Congress Cataloging-in-Publication Data

Baars, Donald.
Navajo country: a geology and natural history of the
four corners region. / Donald Baars.—1st ed. p. cm.
Includes bibliographical references and index.
ISBN 0–8263–1587–9
1. Geology—Colorado Plateau. I. Title.
QE79.5.B285 1995
557.91'3—dc20
95–4344 CIP

CONTENTS

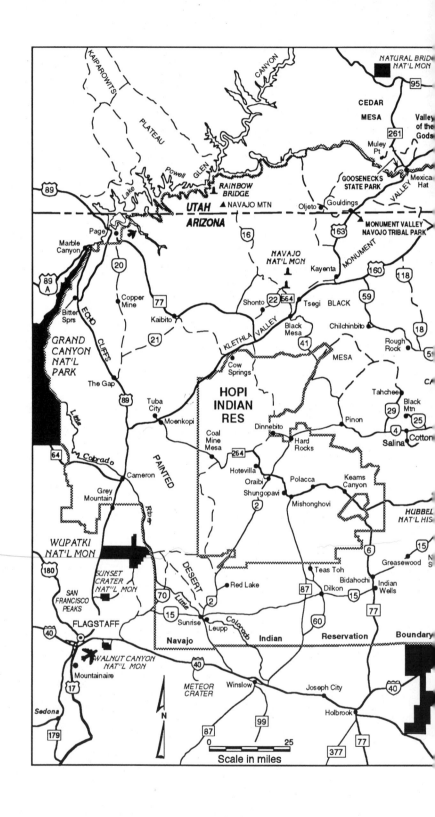

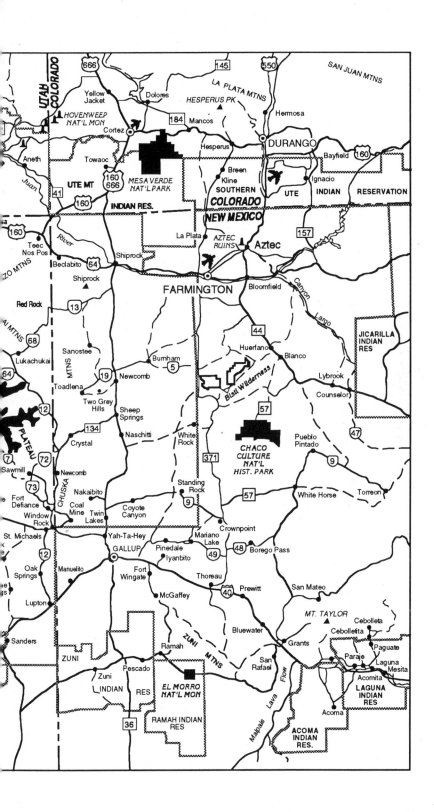

FOREWORD

Beauty. Harmony. These words encapsulate the Navajo way. Living in harmony in the beauty of nature is the essence of Navajo religion. No Anglo religion could better prepare one to survive in Navajo Country—a harsh but beautiful environment. This is *Diné Bikéyah*: NavajoLand, nestled in what is now the Four Corners region, between four sacred mountains. It is a land of contrasts, chosen by the *Diné* (Navajo word for themselves "The People"), respected by the *Diné*, loved by the *Diné*! It is a land of plenty for those who understand it, for those who are survivors, for those with a rich cultural heritage.

Diné Bikéyah is a land of rock, little water, sparse vegetation, and beauty: the southern Colorado Plateau Province. Surrounded by the chaos of upheaved and distorted rocks, the Colorado Plateau is the result of a relatively peaceful geologic history. Perhaps that is why the *Diné* settled here, at least 500 years ago. Perhaps that is why the five-fingered people arose here from the fourth underworld, to survive against all odds of nature. Perhaps Ever Changing Woman and the Holy People knew more than we give them credit for—more of the beauty and tranquility of nature on the Colorado Plateau than we ourselves realize today.

Diné Bikéyah, NavajoLand, is carefully defined by those who live here. The eastern boundary is along the Rio Grande, guarded by Blanca Peak, known as the holy mountain called *Sis Naajini*. The southern limits are marked by Mount Taylor, or *Tsoodzil*. To the west, the San Francisco Peaks, *Dook' o' oosliid*, near Flagstaff, Arizona, and the course of the Colorado River separate *Diné Bikéyah* from Paiute territory, and to the north, Hesperus Peak in the La Plata Range, called *Dibé Nitsaa*, and the San Juan River distinguish *Diné Bikéyah* from Ute territory. The four sacred mountains, most borne of fire, are geographic sentinels, beacons for human navigation in this vast region. Of course, the wisdom of the Anglo invaders caused lines to be drawn on maps to define the Navajo Indian Reservation, but these are not nearly so apparent on the ground as the sacred mountains.

Before A.D. 1500, perhaps long before, the *Diné* settled here, coming piecemeal from hostile country far to the north in what is today western Canada; linguistics establish this pattern of dispersal. Old NavajoLand *(Dinétah)* was the region running roughly from Shiprock on the west, along a strip south of the San Juan River, through where Farmington and Aztec are today, perhaps as far east as present-day Pagosa Springs and Chama. The territory of the farming Apaches gradually spread with time to the limits defined by the four sacred mountains. There were no legal boundaries to these lands until the treaty of 1868, when Congress established the formal Navajo Indian Reservation to include only the vicinity of the Chuska Mountains. Since that time, the *Diné* have dramatically increased in numbers and cultural prosperity, and the Reservation has been expanded accordingly to its present limits largely by administrative decrees. Still, Old NavajoLand is only partially included as "checkerboard" allotted lands; most of the Reservation lies to the south and west of the historic homeland.

It is important to understand the natural relationship between the character of the lands and the present-day Navajo way of life. To this end, it is necessary to examine carefully the origins of the physical landscapes and the geological history that has led to relative Navajo prosperity in our changing world. The two go hand in hand, more so than has ever been realized.

A few words of explanation are in order to elucidate the essence of this book. My present-day address in Kansas may cast doubts on my authenticity and qualifications writing.

I first saw Navajo Country in the early summer of 1952 after graduating from the University of Utah. I was beginning my professional career as a petroleum geologist for Shell Oil Company. I was part of a team of field geologists assigned to map the surface geology of the northern Navajo Indian Reservation from about the Arizona-Utah state line northward to the San Juan River. This exercise was followed by sporadic drilling by the company to determine the petroleum potential of the region, leading to discoveries of major accumulations of oil and gas. During this drilling phase, I was assigned well-sitting chores on most of the wells drilled on and near the Reservation.

Since this dubious beginning, my geologic career has revolved around this fascinating region, and I have returned again and again for one reason or another. I became a relentless "river rat," making innumerable boat trips down the San Juan and Colorado rivers, even becoming a professional river guide and outfitter at one point. My thirty-five-year career studying the petroleum geology of the Paradox Basin always hinged upon these interests in Navajo Country just to the south. Later, I made a regional study of the vast checkerboard acreage along the southern Reservation boundaries held by the Santa Fe Energy Company, an offspring of the Santa Fe Railroad, which acquired the acreage in exchange for establishing an historical railroad route in the 1800s. I served as an expert witness for the United States Justice Department on a major lawsuit initiated by the Navajo Nation regarding drilling and production practices on the Reservation between 1920 and 1946. Finally, I contracted numerous prospective drilling locations on the Reservation with the Chuska Energy Company, which is presently exploring Navajo lands in a second-generation effort to increase Navajo petroleum production. Thus, my professional and personal interests have involved Navajo Country for more than forty years.

Several people deserve my sincere thanks for tolerating my activities during the preparation of this summary volume. Certainly not the least of my appreciation is due to members of the Kansas Geological Survey, with whom I have been employed since 1988; Lee C. Gerhard, Director and Kansas State Geologist has been very patient. Renate Hensiek bravely saw the computer drafting of the many illustrations to completion, and John Charlton was responsible for endless photographic processing; their efforts have contributed greatly to the visual success of the book.

The magnificent low-level aerial color photographs reproduced here were taken by Adriel R. Heisey, a professional pilot for the Navajo Nation and professional photographer in his off time. Adriel built his light plane for photographic purposes over a 15-month period, and transports the folding aircraft by trailer between project sites, using dirt roads for runways. He flies amid spectacular buttes and towering spires at 40 miles per hour, flying the aircraft by strapping the control stick to his right knee; he adds ". . . my own feet [and the rudder pedals] are the only things between me and all creation." Using two complete camera systems on each flight, based on Pentax 6 x 7 cm cameras, his helicopter-like photographic platform provides unique and intimate vantage points rivalled by no other technique. I am greatly indebted to Adriel for permitting the use in this book of a mere sampling of his many outstanding photographs.

The Navajo Nation, with the welcome assistance of Brad Nesemeier of its Minerals Department, kindly permitted my photographic and geological review escapades back into Navajo Country in 1992. Brad also reviewed an early draft of the manuscript and made valuable suggestions. (Any persons wishing to conduct geologic investigations on the Navajo Reservation must first apply for and receive a permit from the Navajo Nation Minerals Department, P.O. Box 146, Window Rock, Arizona 86515.)

Finally, the deeply appreciated encouragement and patience of my wife Jane have made this book possible.

NAVAJO COUNTRY

GEOLOGIC TIME SCALE

ERA	PERIOD	MILLIONS OF YEARS AGO*	FOUR CORNERS FORMATIONS
CENOZOIC	Quaternary	0–1.6	soil, sand, gravel
	Tertiary	1.6–65	Terrace gravels, Diatremes, Igneous intrusives
MESOZOIC	Cretaceous	65–135	Mesaverde Group, Mancos Shale, Dakota Sandstone, Cedar Mtn./Burro Canyon
	Jurassic	135–205	Morrison Formation, Entrada Sandstone, Carmel/Wanakah Formation, Navajo Sandstone, Kayenta Formation, Moenave Formation, Wingate Sandstone
	Triassic	205–250	Chinle Formation, Moenkopi Formation
PALEOZOIC	Permian	250–290	Cutler Group: DeChelly Sandstone, Organ Rock Shale, Cedar Mesa Sandstone
	Pennsylvanian	290–325	Halgaito Shale, Hermosa Group: Honaker Trail Fm., Paradox Formation
	(Rocks not exposed)		Pinkerton Trail Fm., Molas Formation
	Mississippian	325–355	Leadville/Redwall Fm.
	Devonian	355–410	Ouray Limestone, Elbert Formation
	Silurian	410–438	Rocks missing
	Ordovician	438–510	Rocks missing
	Cambrian	510–570	Ignacio-Lynch Fms.
PRECAMBRIAN	Upper	570–2,500	Quartzite/granite
	Lower	2,500–4,500?	Metamorphic/granite

{*Dates in millions of years from Cowie and Bassett, 1989}

Geologic Time Scale. This scheme was devised largely in Europe but widely recognized in North America, showing commonly used formation names in the Four Corners area of Navajo Country. Ages in millions of years vary widely for different authors, but those shown are from an international system published by Cowie and Bassett in 1989.

COLORADO PLATEAU PROVINCE

G reater Navajo Country comprises approximately the southern one-third of the Colorado Plateau Province. Not much of the Colorado Plateau actually lies within the state of Colorado. Rather, it constitutes the Colorado River drainage basin. The Province is generally a land of high plateaus, separated from the Midcontinent by mountainous regions consisting of geologically uplifted blocks and intervening down-dropped basins. It is separated from the western and southern Basin and Range provinces by prominent escarpments that represent ancient fault zones, obscured by eons of erosional events.

Although not prominent features at the surface of the Colorado Plateau, this book will be dealing with faults in nearly all of its discussions about the origin of the Colorado Plateau in general, and Navajo Country in particular. A fault, in geological terms, is a fracture in the Earth's crust along which movement has taken place. This movement may be vertical, in which case one block of rock has been lifted or dropped relative to an adjacent one, or horizontal, when one block has move to the left or right of its counterpart. Faults with minor amounts of movement are omnipresent in rocks at the Earth's surface, but those of importance to understanding the Colorado Plateau and Navajo Country are major features that are rarely exposed yet must be assumed to underlie the Province boundaries, control the significant uplifts, and localize the great basins of sedimentary deposition.

Geologically speaking, the Colorado Plateau Province is separated from neighboring parts of the continent by ancient faults, or zones of faulting, that originated between 1 and 2 billion years ago in the Precambrian Era. These ancient faults are of continental scale, separating the continent into orthogonal blocks that measure 250 miles or more across and, at least in Precambrian time, were faults of the horizontal variety, termed "wrench faults." The Colorado Plateau, bounded by these ancient wrench faults, is but one of many fault-bounded blocks that comprise the continent. It is unknown how, or why, these huge fault blocks formed, but they constitute Earth architecture on every continent. They all formed at about the same time, namely 1.7–1.6 billion years ago. They are invariably oriented northwest to southeast and northeast to southwest forming diamond-shaped provinces, such as our Colorado Plateau.

The Colorado Plateau is distinguished from surrounding country by its relatively simple geologic structure. Adjacent provinces are typified by fault-bounded uplifts separated by sharply down-dropped basins in apparent chaotic disarray, forming mountain range after mountain range. The Plateau country, on the other hand, consists of relatively flat-lying rocks, with only rolling structural features that produced vast range lands separated by rounded uplands. The entire Province is high in elevation, and drainage has cut magnificent, unique canyons on its way to the sea.

The difference seems to be that the Colorado Plateau, isolated by its bounding fault systems, has acted as a rigid block of the Earth's crust, surrounded by other fault blocks that were fractured like brittle glass in later structural events. The crust of the Earth has been contorted, folded and faulted in numerous mountain-building episodes called "orogenies," but the Colorado Plateau only bent and folded over preexisting structures near the end. Some say it is because the Earth's crust is thicker here; whatever the reason, it has acted independently of the rest of the continent.

MONOCLINES

The geology of the Colorado Plateau is typified by a series of folds in the rocks of gigantic proportions. The upfolds are of a special kind that are called monoclines. Ordinary upfolds in layered rocks, called anticlines, are symmetrical in cross section, having rock layers that dip away from a crest much like a fold formed in a piece of paper when the two edges are pushed together. A monocline, however, is only half, one flank, of an anticline. The rock layers are flat on top, folding down to another lower flat surface, like a carpet draping across a stair step. This, in fact, is almost how they form. Basement rocks beneath the higher layered rocks have been faulted, forming the "stair step," and the layered rocks above have draped across the step (see figure 1). To make matters even worse, these monoclines on the Colorado Plateau are not playing fair. The rocks are neither flat above or below the sharply draping fold but instead dip gently away from the top of the fold in one direction and rise gradually away from the down-side of the fold in the

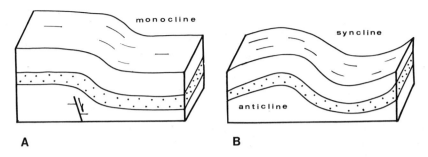

A   B

FIGURE 1 Simplified Block Diagrams. These two figures show differences between a monocline (A), in which sedimentary layers drape across basement faulting, and simple upfolds, anticlines, and downfolds, synclines (B) caused by lateral compression in the Earth's crust.

other. In reality, they are huge asymmetrical anticlines, but the folds are so sharp as to appear as monoclines when viewed from the ground.

Examples of Colorado Plateau structures bounded by monoclines are the San Rafael Swell south of Price, Utah, the Monument Upwarp that lies between Blanding or Monticello and Hanksville, Utah, the East Kaibab monocline that forms the eastern limit of the Grand Canyon, and the Defiance and Hogback monoclines that trend along the Arizona–New Mexico border south of the Four Corners. All of these are generally north-trending, and in all the layered rocks drape sharply down to the east (see figures 2 and 3). A few, such as the abrupt fold along the western margin of the Nacimiento–San Pedro Mountains north of Albuquerque, New Mexico, drop the layered rocks down toward the west.

Magnificent canyons bisect two of these classic monoclines. The Grand Canyon has been carved by the Colorado River across the East Kaibab monocline, and the San Juan River has cut canyons directly across the girth of the Monument Upwarp. Of these, Grand Canyon is now so deep as to expose the faulting beneath the fold, showing that these features are indeed layered rocks draping over stair-step fault blocks. Basement faults are not seen at the surface beneath the other monoclines, but geophysical studies and deep drilling indicate that they are present at depth.

BASINS

Basins lie between these peculiar monoclines. Broadly downfolded rocks rise gradually from the bottom of one monocline until they rise to the crest of the next. These regions where the downfolded rocks are prevalent are so vast that they appear as great, unpretentious plains. It is only the monoclines that distinguish the basins that are strikingly noticeable features (see figure 2). However, it is the basins, the dull gray broad and flat countrysides of the Colorado Plateau, that are of economic significance.

It is common knowledge that the high country is more thoroughly and deeply eroded than the low country. Storms attack the most fiercely, runoff is greatest, and erosion is deepest in high places. Thus, the upfolds have been eroded down deeper into the older layered rocks, and higher, younger rocks are best preserved in the basins, or downfolded regions. It is in these younger sedimentary rocks that oil, gas, and coal occur. The uplifted regions may have their beauty, their deep canyons, their red rocks, their scenic mesas, their Grand Canyons, their Canyonlands, their Monument Valleys, but the dull, gray basins are where the wealth lies.

As an indication of the scale of these features, the upfold known as the Monument Upwarp is about 80 miles long and 30–40 miles across, depending upon how its flanks are defined. However, the highways crossing the east-flanking monocline, Comb Ridge, traverse the sharp fold in a mile or two. The San Juan Basin that lies between the Defiance and Nacimiento uplifts is 100 miles or more across, depending upon where it is measured. These structures are so broad and huge as to be unrecognizable to the casual observer. It is only when they are marked on maps that their existence is realized, or when some geologist points them out.

FIRE MOUNTAINS

A variety of landforms result from the presence of once-molten rocks. The term "igneous" is used for rocks formed from melted Earth materials. Of course, all kinds of igneous rocks originated at considerable depths in the Earth, and it is the manner in which they came to be seen at the Earth's surface that distinguishes the various types. The most commonly recognized of these are "extrusive" igneous rocks, for as the name implies, the molten matter—lava or ash—was extruded onto the surface from volcanoes of one kind or another. Hawaiian volcanoes are familiar examples. The more prominent of the volcanoes on the Colorado Plateau are Navajo sacred mountains: Mount Taylor near Grants, New Mexico, and San Francisco Peak near Flagstaff, Arizona. These

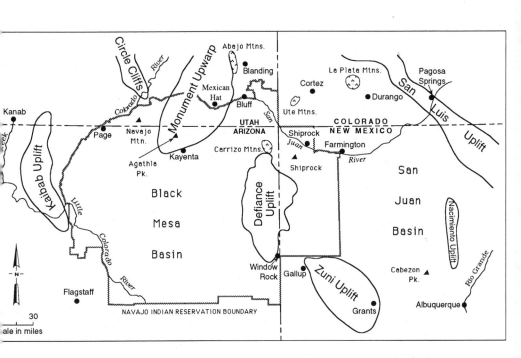

FIGURE 2 Map of Major Structural Features in Navajo Country. The named uplifts are structurally high features where older rocks of Paleozoic and Precambrian age are exposed at the surface today. The two large basins, the San Juan and Black Mesa, are structurally low areas where younger layered rocks have been preserved from erosional removal. Oval areas containing inverted "v" patterns are prominent exposures of intrusive igneous rocks (laccolithic ranges). Outlines of these major features appear on maps presented later for geologic and geographic orientation.

and closely related volcanic features have been actively eruptive in nearly recent times, as shown by their very fresh, almost unweathered characteristics.

Another variety of extrusive igneous rock that typifies the Colorado Plateau landscape is seen at the necks, or vents, of highly explosive, gaseous eruptions. In these, massive explosions of deep-seated gases ripped their way to the surface, something like the historic eruption of Krakatoa, bringing up small to very large fragments of rock, through which the gases travelled, expelling them violently onto the surface. These rock fragments, consisting of high-temperature rock types from the Earth's mantle along with pieces of granite, metamorphic basement rocks, and various sedimentary rock layers, were blown out at the surface, some falling back into the vent to harden. Such volcanic vents are termed "diatremes" and form such prominent topographic features as Shiprock near the Four Corners and Agathla (or El Capitan) Peak south of Monument Valley (see figure 2). There are hundreds of lesser-known diatremes in Navajo Country.

Other types of igneous occurrences are called "intrusive" igneous rocks because the molten matter, or magma, cooled and hardened into rock without reaching the surface. These are crystalline, more coarse-grained igneous rocks, formed by a much slower cooling process at depth. The molten bodies either forced their way between other preexisting rocks, or simply melted and incorporated a host rock as the process progressed. Large rock bodies of this type form prominent mountain ranges on the Colorado Plateau, where erosion has exposed them at the surface. Irregularly shaped intrusive bodies, such as stocks, dikes and sills, intruded locally into the existing sedimentary rock cover, forming complex igneous bodies, generally, but not exactly correctly, called "laccoliths." By definition, a laccolith is a mushroom-shaped intrusion, having risen along a tubelike fissure and then injected outward between bulged layers of sedimentary rocks. Some of the now-exposed bodies of this shape are seen, and it was originally believed that most of the intrusive bodies were of this pattern. Thus, the several prominent mountain ranges on the Colorado Plateau of this origin are generally called laccolithic ranges. Examples are the La Sal Mountains near Moab, Utah, the Ute Mountains (or Sleeping Ute) near Cortez, Colorado, and the Carrizo Mountains near the Four Corners (see figure 2). Another of the Navajo sacred mountains, Mount Hesperus in the La Plata Mountains, is laccolithic in origin. Because the intrusive igneous bodies are not mushroom-shaped, but instead have a three-directional pattern, a better descriptive term—"cactolith"—has been applied, suggesting that the intrusives are cactus-shaped. However, that term has not caught the fancy of geologists, and the mountains are still called laccolithic.

These geologic features—gentle uplifts, monoclines, broad basins, diatremes, and laccolithic ranges—form the Colorado Plateau; these provide the framework of *Diné Bikéyah:* Navajo Country.

CHAPTER TWO

NAVAJO LANDSCAPES

The scope of this book extends beyond the boundaries of the formal Navajo Indian Reservation to cover the ill-defined land lying between the four sacred mountains of the Navajo people that are significant. In Navajo tradition, all natural features are discussed according to the four cardinal points of the compass.

EAST

East is always the most important and traditionally the starting point of such discussions because that is the direction from which the sun rises. All Navajo dwellings, the hogans, are built with their entryways pointing eastward. It was from this direction that the Holy People emerged to instruct and create the first five-fingered people of the fifth world. First Man and First Woman, assisted by Black Body and Blue Body, formed the four sacred mountains that are the guarding sentinels of Navajo Country.

The sacred mountain that bounds Navajo Country to the east is generally believed to be Blanca Peak, *Sis Naajini*, although other mountains in the vicinity have been suggested as the official home of Rock Crystal Boy and Rock Crystal Girl who were placed there as guardians by First Man and First Woman. They anchored the mountain with lightning and decorated it with white shell, white lightening, white corn, and dark clouds to produce the sudden and harsh male rains. Indeed, Blanca Peak, or Sierra Blanca as it was called by the Spanish explorers, means "White Mountain."

From a geological perspective, the eastern limits of Navajo Country might best be placed at the Rio Grande that flows past the base of the sacred mountain. Not large as major rivers go, the Rio Grande drains the eastern Colorado Rocky Mountains toward the south along a series of open, irregularly interlocking valleys that were not formed solely by river erosion, but instead followed geologic structure. Geologists call this the Rio Grande Rift, the faulted boundary between the Colorado Plateau on the west and the Midcontinent on the east. Geographers would say that this is a series of north-south valleys dividing the Rocky Mountains into physiographic subdivisions, but geologically speaking, the Rio Grande Rift is a major separation line for distinctive geologic features that date back to Precambrian time, perhaps 2 billion years ago.

Obvious features at the surface are fault-bounded basins of deposition that are offset, but interconnected en echelon, by the course of the river. Resulting topographic features are the San Luis Valley of southern Colorado and the Española and Albuquerque valleys of New Mexico. Sedimentary deposits within the individual basins are very thick and geologically rather young, having been deposited within the past 30 million years or so. These deposits, and the associated recent faulting and volcanic activity along the rift, have attracted the intense interest of geologists for decades. Although it has been realized that the rift is a highly important geologic feature, dividing the very stable continental interior from the more restless crustal blocks to the west, what is not generally considered is the very ancient origins and deep-rooted significance of this segment of the basement structural fabric.

A short distance to the west of the Rio Grande Rift lies a generally parallel, north-south oriented mountain chain called the Nacimiento–San Pedro Uplift. The range consists of a core of ancient Precambrian granitic and metamorphic rocks, covered along its flanks with a thin veneer of sedimentary layered rocks. All indications are that the uplift was originated in Precambrian time and has since been further uplifted a number of times in geologic history. It is bounded on the west by a sharp monoclinal fold that comprises the eastern limits of the San Juan Basin.

SOUTH

First Man and First Woman constructed the sacred mountain to the south, Mount Taylor, or *Tsoodzil*. They used a great stone knife to hold the mountain in place and decorated its summit with turquoise, many different animals—including especially the Blue Bird—and added dark mist to produce the gentle female rain. Then they covered it all with blue sky and stationed Boy Who is Bringing Back Turquoise and Girl Who is Bringing Back Many Ears of Corn to protect the sacred place.

Mount Taylor is a fairly young volcano that lies just to the north of the Zuni Uplift. While Mount Taylor is the most prominent landmark

MOUNT TAYLOR   This breathtaking peak, the sacred mountain of the south, rises above Grants, New Mexico. The upland is a relatively recent volcano.

in the region, the Zuni Uplift is of greater geologic significance. It is a very large upfold that trends northwesterly from southeast of Grants to near Gallup, New Mexico. The fold is asymmetrical, its steepest flank being a monocline on the south that downfolds the layered rocks toward the southwest. It is a very ancient structure, having been present in nearly the same configuration since Precambrian time. The northwesterly trend of the uplift and its sharp flanking monocline on the south indicate that ancient basement faults have controlled its geographic location and shape. The sharp northwesterly plunging nose of the anticlinal fold forms prominent hook-shaped hogback ridges just east and north of Gallup, near Fort Wingate.

Gallup sits in a narrow saddle between the northwest nose of the Zuni Uplift, and the next row of mesas rising to the north into the Chuska Mountains and neighboring Defiance Plateau. The Chuskas and Defiance Plateau, known generally to geologists as the Defiance Uplift, separate the San Juan Basin to the east from the Black Mesa and Holbrook basins to the west.

WEST

More accurately referred to as the southwest, this areas is where First Man and First Woman fastened the mountain we call San Francisco Peak, or *Dook' o' oosliid*, with a sunbeam and adorned the summit with abalone shell. They placed many animals here—including especially the Yellow Warbler—and stationed White Corn Boy and Yellow Corn Girl to oversee the sacred place. They covered the mountain with a dark cloud to produce the violent male rain and shrouded all of this with a yellow cloud.

The San Francisco Peaks, like Mount Taylor, are a fairly recently active group of volcanoes. They rise prominently above the southeastern corner of what geologists call the Kaibab Uplift, the eastern border of which is the East Kaibab monocline along with other closely related structures. It is through this high plateau country to the north of San Francisco Peak that the Colorado River has carved Grand Canyon. There it can be clearly seen that the monocline consists of layered sedimentary rocks that drape across a Precambrian fault block.

Another more or less northerly-trending monocline, known as Echo Cliffs, parallels the East Kaibab monocline to the east, forming a row of cliffs between Cameron and Lees Ferry, Arizona. Layered rocks first rise sharply toward the west from Black Mesa Basin along the Echo Cliffs monocline, and the harder, older (Paleozoic) rocks form a plateau through which the Colorado River has eroded Marble Canyon on its way into Grand Canyon, and the Little Colorado River has cut a deep gorge to join the Colorado River. There, the combined rivers cross the East Kaibab monocline, which brings still older—Precambrian—rocks upward to river level to complete the stair-step rise of the older sequences of rocks into full view.

NORTH

Mount Hesperus, or *Dibé Nitsaa*, in the La Plata Mountains of southwestern Colorado, is the sacred mountain of the north. Here First Man and First Woman bolted the mountain down with a rainbow and decorated its summit with black beads of jet. They placed there many different plants and animals, covered it with the gray clouds that form gentle female rains, and then covered it all with darkness. They placed Pollen Boy and Grasshopper Girl to reign over the dark sky country, also known as the Place of Big Mountain Sheep. It is here that White Shell

LA PLATA MOUNTAINS   The highest point, Mount Hesperus (left), rises majestically above Durango, Colorado. This peak, the sacred mountain of the north, consists of laccolithic igneous rocks that have been intruded into sedimentary rocks of Paleozoic and Mesozoic age. Dark-colored badlands of the Mancos Shale, capped by Point Lookout Sandstone of Cretaceous age, comprise the middle ground beyond the city in this aerial view.

Woman found her final resting place. In Navajo mythology, north is the cardinal compass direction that signifies darkness, or black colors, and it is the direction of fear and ghostly events. When a Navajo who has not attained a ripe old age dies in a hogan, the door and smoke hole are sealed, and the north wall of the hogan is broken out to let the ghosts and spirit of the dead person escape to the north.

The La Plata Mountains, their name meaning "silver" in Spanish, are formed of a complex of stocks, dikes, sills, and laccoliths of igneous rocks intruded into sedimentary layered rocks just southwest of the San Juan Mountains. The La Platas lie just north of the Hogback monocline that forms the northwestern margin of the San Juan Basin, extending northeastward from near Shiprock to a short distance north of Durango, Colorado. Geophysical studies and deep drilling show that a basement fault with large vertical displacement underlies the Hogback

GOBERNADOR KNOB   The birthplace of Changing Woman of Navajo legend and one of the inner sacred mountains, Gobernador Knob consists of sedimentary rocks of Tertiary age in the heart of the San Juan Basin south of Bloomfield, New Mexico.

monocline. Another monoclinal fold extends from Durango to near Pagosa Springs, Colorado, completing the northern limits of the San Juan Basin.

LANDS BETWEEN THE SACRED MOUNTAINS

To set the stage for the geology of Navajo Country and to summarize this section, it should be made clear that there are two large basins—the San Juan Basin on the east and the Black Mesa–Holbrook Basin complex on the west—each bounded by uplifts of very ancient origins. The geologic backbone of Navajo Country is made up of the Chuska Mountains and Defiance Plateau (also known as the Defiance Uplift) that separate the two basins. The San Juan Basin is bounded on the east by monoclines and underlying basement faults of the Chama and Nacimiento–San Pedro Mountains, on the south by the Zuni Uplift, on

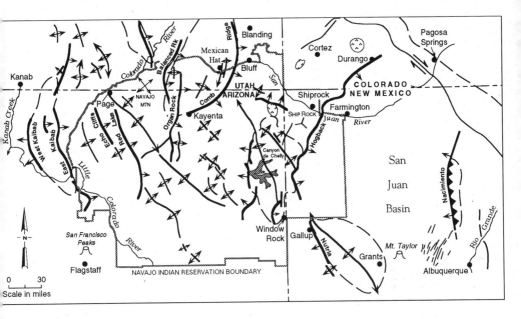

FIGURE 3 Map of Major Anticlines and Monoclines in Navajo Country. Monoclines are represented by heavy lines with arrows pointing toward the downside of the fold and are labelled as to their names; the Nacimiento monocline in the southeastern area of the map is shown as being thrust-faulted. Anticlines are indicated by lighter lines at their crests with double arrows showing strata that are folded down in both directions away from the axis; intervening synclines have been omitted for clarity. Larger volcanic features, Mt. Taylor and the San Francisco Peaks are shown diagramatically. These and Mt. Hesperus in the La Plata Mountains constitute three of the four sacred mountains; Blanco Peak, the fourth, lies to the northeast of the map area. Outlines of major uplifts (see figure 2) are shown by dashed lines.

the west by the Defiance Uplift, and to the ominous north by the Hogback monocline and La Plata Mountains. West of the Defiance Uplift (Chuska Mountains) is the Black Mesa–Holbrook Basin complex, bounded on the south by the faulted escarpment of the Mogollon Rim, to the west by the Echo Cliffs and East Kaibab monoclines, and to the north by the Comb Ridge monocline of the Monument Upwarp (see figure 3). All this is guarded by four sacred mountains.

CHAPTER THREE

IT'S ABOUT TIME

The concept of geologic time is practically beyond human comprehension. We tend to think of time in terms of only a few days, weeks, years, or even decades, for these constitute a lifetime. The events of a few hundred years ago are considered ancient history, and it is almost impossible for us to conceive of a time a million, a hundred million, or (heaven help us!) a billion years before now. Yet the geologic history of Navajo Country goes back more than 1.8 billion years!

Geologists are not much better than ordinary folks in this matter, so they devised the Geologic Time Scale. With this scheme, they could put chunks of time into handy boxes for easier reference and name the time boxes for pretty places where they enjoyed studying particular layers of rocks. Each box, a geologic time period, represented the time it took to deposit one's favorite bed of layered rocks—whatever length of time that might be.

Early attempts to establish a system of classifying spans of time were not very successful. For example, Johann Gottlieb Lehmann proposed, in 1766, that mountain-building episodes be classified into (1) those that formed at the time of the creation of the Earth, (2) those that formed during the time of "The Flood," and (3) those that formed since "The Flood." However, there have been hundreds of floods during the course of geologic time, so that scheme didn't work well.

By the 1830s, the scheme we now use began to emerge from the work of a few British geologists, but not without some frustration. Adam Sedgwick studied and published descriptions of the oldest sedimentary rock sequence in the pleasant countryside in Wales, naming it the CAMBRIAN System, Cambria being the Latin name for Wales. Simultaneously, Sir Roderick Murchison studied the sedimentary rocks along the border between England and Wales, naming his sequence the SILURIAN System, the Silures having been early inhabitants of the region. Both names were published in 1835. As luck would have it, some of the rocks included in each system were the same strata; the two systems overlapped. Of course, a professional skirmish broke out. It continued until 1879, when Charles Lapworth settled the matter by naming the overlapping section the ORDOVICIAN System for another ancient tribe of people, the Ordovices. Murchison and Sedgwick, in 1839, had named the next younger geologic period the DEVONIAN for Devonshire, a lovely summer resort area in southwestern England.

The next younger sequence of rocks in the region is the CARBONIFEROUS System, the coal measures of central England. The actual coal-bearing beds are in the upper part, overlying a massive cliff-forming limestone known throughout western Europe as the Mountain Limestone. Although of totally different origin and surface appearance, the two rock types are still called the "Lower" and "Upper" Carboniferous in Europe. An almost identical sequence occurs in the United States as well, where it was subdivided into the older (and therefore lower) MISSISSIPPIAN System and the younger PENNSYLVANIAN System in 1869 and 1891, respectively. We use the two names effectively in this country, but European geologists still refuse to recognize the only American names in the Geologic Time Scale.

By this time, Sir Roderick Murchison had become obsessed with naming all rocks for use in an international Geologic Time Scale. He knew that an unnamed series of rocks existed between the Carboniferous System of England and the much younger Triassic System of western Europe, but he was at a loss for an appropriate name. He somehow wangled an invitation from Czar Nicholas I to visit Russia during the summers of 1840 and 1841 under the guise of determining whether his other named systems of England and Wales were useful in eastern Europe. He travelled extensively into the Ural Mountains, accompanied by the Czar's wife and a team of Russian geologists, where he found the missing section rather well exposed. In an 1841 report to the Czar, he named the section of mostly red beds and evaporites the PERMIAN System for the city of Perm and the extensive Perm Basin. Unfortunately, the section Murchison studied contains few distinctive fossils, and the lower boundary of the Permian System has only recently (August 1991) been established by Russian geologists.

At first, the "absolute" ages of rocks were unknown, and the named boxes of geologic time were shuffled into order based on their RELATIVE position in the overall sequence of rocks. Thus, the oldest rocks were considered to be those at the bottom of the stack, as they must already have been in place before the next overlying layer could be deposited on top. This idea has become known as the Law of Superposition. Once the original order of the layered rocks was established, the contained fossils and their sequence of evolutionary changes were utilized extensively to date and correlate sedimentary rocks on a global scale in a relative sense. During the intervening 150 years, methods have been developed for measuring the age of rocks, in years, by studying radioactive decay rates.

The named geologic periods previously discussed are lumped into a larger unit, the Paleozoic Era (from the Greek *palaios*, meaning "old" and *zoe*, meaning "life" or "ancient life"). Paleozoic time was preceded by the Precambrian Era, the first 4 billion years or so of Earth history. No universally acceptable subdivision of the very lengthy Precambrian Era has ever been proposed despite numerous attempts. The Paleozoic Era was followed by the Mesozoic ("middle life") and that, in turn, by the Cenozoic ("recent life"), both names derived from Greek.

The Mesozoic Era is subdivided into three geologic periods: the TRIASSIC Period, named for a tripartite sequence of rocks in Germany; the JURASSIC Period, named for rocks of the Jura Alps along the French-Swiss border; and the CRETACEOUS Period, named for the chalk cliffs in the Paris basin (from the Latin *creta*, meaning "chalk").

The Cenozoic Era is subdivided into two periods—Tertiary and Quaternary—the names being carry-overs from the first known Geologic Time Scale published in 1760 by Giovanni Arduino. He designated three periods of which Tertiary (third) stuck in the present-day scale. It was soon realized that younger sedimentary deposits occur above rocks considered to be Tertiary in age, and a Frenchman, Paul G. Desnoyers, following the original scheme, classified these as QUATERNARY (fourth) in age in 1829. These names remain in common usage today.

ROCKS ARE ROCKS ARE ROCKS

There are three broad categories of rocks: igneous, metamorphic, and sedimentary. Each has a different story to tell.

IGNEOUS ROCK forms as molten bodies cool and crystallize. If the molten mass cools beneath the Earth's surface, it is called an intrusive igneous rock; a common example is granite. If hot liquids are poured out at the surface from a volcano, either as lava or ash, an extrusive igneous rock results.

METAMORPHIC ROCK forms as a result of the alteration of other rock types by intense heat and pressure deep within the Earth. Minerals in the rock are distorted and/or new minerals are formed by partial melting and recrystallization. Resulting rock textures are mashed or wildly contorted and contain strange minerals such as mica or garnet. Metamorphic rocks are usually very old.

SEDIMENTARY ROCKS contain compacted and/or cemented sediments, such as pebbles, sand, and mud, derived from the weathering of any other rock. Pebble deposits become conglomerate, sand is cemented to form sandstone, and mud compacts to shale. Limestone consists of calcium carbonate sediments, most of which are derived from the hard parts of marine plants and animals. They may occur as mud-size sediments or as sand- or pebble-size fragments of shell material, and fossils of complete and broken shells are common. Most limestones were deposited in warm, shallow seas, where organic activity is high.

A special variety of sedimentary rocks forms by direct precipitation of salts from concentrated seawater and highly saline lakes, such as the Great Salt Lake in Utah. Because intense concentration of various salts by extremes of evaporation of the water causes the salts to precipitate from solution, the resulting deposits are known as evaporites. A common rock that forms in this way is gypsum, consisting of hydrous calcium sulfate ($CaSO_42H_2O$), a soft, usually white and chalky mineral, and various crystalline forms known as selenite, alabaster, or satin spar. The sulfates are the least soluble of the common natural salts and consequently are among the first minerals to form by evaporation processes. When gypsum loses its water molecules due to deep burial or heightened temperatures, it becomes a harder, more dense rock called anhydrite (meaning "without water"). Anhydrite is usually only found in buried rocks, as it absorbs moisture easily and reverts to gypsum upon weathering at the surface. Another common evaporite often found in association with gypsum or anhydrite is halite, best known as table salt, consisting of crystalline sodium chloride ($NaCl$). Halite precipitates from natural solution following gypsum, as it is more soluble in water. Finally, relatively rare salts of potassium form bitter-tasting rocks, known as sylvite (KCl), and other unusual minerals of similar composition may be precipitated from solution. These rocks are common and locally thick in the deep subsurface of the Paradox Basin north of Navajo Country.

Rocks of all kinds are found in Navajo Country. Igneous and metamorphic rocks, being more resistant to erosion, occur in the highest peaks and ranges in the San Juan Mountains and in the dark depths of Grand Canyon. Erosion has removed most of the sedimentary rock cover from the San Juan Mountains dome, exposing these harder, older rocks in the core. Erosional remnants of thick sedimentary rocks may be seen dipping away from the uplift in canyons carved from the flanks of the dome. A 16,000-foot-thick section of sedimentary rocks of Cambrian through Tertiary age has been exposed in the Animas Valley immediately north of Durango. Although rocks younger than Paleozoic have been stripped by erosion from the uplifted region of Grand and Marble canyons, the river has been able to cut sufficiently deep in Grand Canyon to expose the metamorphic basement. Extrusive igneous rocks of very recent volcanic activity dot the landscape of Navajo Country.

KEEPING TRACK

Each layer of sedimentary rock has its own topographic expression, color, and mood. Any particular layer is different from those above and below, and it can be traced laterally for miles or hundreds of miles across the country or through deep wells. Each layer of rock has a different geologic history and meaning: one is a stream deposit, another is an ancient field of windblown dunes, still another was deposited in shallow tropical seas. Because of this, it is important for geologists to distinguish between specific layers, and so we give each a unique name.

Most people think of a silly-looking rock, perhaps one shaped like an owl or an elephant, as a rock formation. To a geologist, however, a "FORMATION" is a layer—or series of similar layers—of rock that is geographically extensive, geologically significant, or both. Formation names are usually derived from a location, or "type section," in which the layer(s) can best be studied. The formal name has two parts: the first part is a geographic place-name, usually related to the type section, and the second part designates the kind of rock that typifies the formation. For example, the gray slopes surrounding Durango are weathered from a very thick layer of marine gray shale named the Mancos Shale. The name "Mancos" comes from the type section along the Mancos River valley that lies between Durango and Cortez, and the term "Shale" explains the most dominant rock type. The Mancos Shale is present over several states. It underlies the towns of Mancos, Cortez and Grand Junction, besides Durango, and extends across the Rocky Mountains. The overlying tan cliff, high to the west of Durango, is called the Point Lookout Sandstone, named for the prominent point at the entrance to Mesa Verde National Park and the fact that it consists of sandstone.

Sometimes a whole series of different but closely related rock types occur in considerable thicknesses, making it impractical and unnecessary to name each bed. A series of interbedded red sandstones and shales occurs in the Animas Valley just north of Durango. Since the 2,500-foot-thick pile of red beds cannot be called either a sandstone or a shale, it is lumped into a formation, in this case the Cutler Formation. The name "Cutler" comes from Cutler Creek just north of Ouray, where the same beds are well exposed; the name "Formation" means that it consists of more than one rock type.

Sometimes a formation is given a name, and then it is learned that the rocks are too complex and need to be subdivided into more, smaller units. In this case a formation can be elevated in rank to a "group" that can be divided into two or more formations. An example is the Hermosa Formation, named for rocks in Hermosa Mountain north of Durango, but later found to change elsewhere to different kinds of rocks. The formation was upgraded to the Hermosa Group, with three formations: the lower Pinkerton Trail Formation, the middle Paradox Formation, and the upper Honaker Trail Formation. Alternatively, a formation may be subdivided into "members," which may or may not be given individual names. An example is the Elbert Formation, named for Elbert Creek near Purgatory Ski Area in the San Juan Mountains. In this case, the formation consists of a basal sandstone, named the McCracken Sandstone Member of the Elbert Formation, and an unnamed upper shale section. (Notice how long and complex a member name can be.)

All of this is designed to bring order out of chaos, and it USUALLY works. The trouble is that a single layer of rock is often given different names when studied in different areas. When this happens, the name first applied should be the one used. However, geologists are generally too stubborn and too provincial to follow the rule, and two names often result for the same rock layer. For example, limestone in the low gray cliffs seen around Tamarron Resort north of Durango is called the Leadville Limestone in Colorado, but it is the same bed that crops out in Grand Canyon where it is called the Redwall Limestone. Since geologists at neither end will compromise, the same rock layer has two names.

The art of naming, correlating, studying, and keeping track of all this is called stratigraphy, and those who worry about it for a living are called stratigraphers.

CHAPTER FOUR

DINÉ BAHANÉ
THE CREATION STORY

L ong before First Man and First Woman were created by the Holy
People from the east, long even before the time of the first underground black world where all the people were various insects
known as the Air-Spirit People, Earth was formed. It has been calculated that it was about 4.5 billion years ago when this mass of lumpy
planet congealed, and it has been guessed that the Earth formed over
millions of years as meteors, asteroids, and other forms of space
debris—attracted one to another by gravitational forces—accreted to
form an ever-enlarging mass. It was soon caught up in the orbit of the
nearest star (we call it Sun), and there it has been stuck ever since.
Because no one really knows how this all happened, and no one ever
will know for sure, our tale will start at this point in history.

About 1.8 billion years ago in Navajo Country, the land lay as barren as the back of a horny toad. There was not a sprig of anything
green to be seen anywhere and not a critter to be found—certainly no
gnats. The land was as flat as a pool table, having been thoroughly
eroded to that form on highly metamorphosed gneisses, schists, and
some granite. This was the surface that park rangers in Grand Canyon
call "The Great Unconformity." The rock is what geologists call "the
basement."

Metamorphic rocks are formed when preexisting rocks are subjected to great heat or pressure, or both. The heat may come from
radioactive decay deep within the Earth, or from friction caused by
pressure from the weight of thick overlying rocks. Intense lateral pressure can be caused by tremendous mountain-building stress in the crust
of the Earth. Pressure distorts the newly formed minerals into wavy,
gnarled fabrics forming rocks called gneiss (pronounced "nice"), or
smashes them into platy, papery sheeting to form rocks called schist
(pronounced "shist"). If the process is carried too far and the preexisting rocks are completely melted down, the resulting rock will resemble

granitic rocks. Remember that these rocks are altered preexisting ones so they are not the original rocks of the Earth's crust.

This metamorphic basement is only exposed in a few places in greater Navajo Country. To the east, basement rocks are exposed in the Nacimiento Mountains; to the south they are found in the core of the Zuni Mountains; to the west they crop out in 1,000-foot-high canyon walls in Granite Gorge in Grand Canyon where the rock body is called the Vishnu Schist; and to the north the Twilight Gneiss in the high San Juan Mountains consists of a similar metamorphic complex. Where these basement rocks have been dated by means of studying rates of decay of their radioactive minerals—in the San Juan Mountains and in Grand Canyon—they date at about 1.7–1.8 billion years old; that is the time the minerals last crystallized. The oldest known rocks in North America, found in the Slave Province of northern Canada, have recently been dated at 3.96 billion years, so there is much of Earth history in Navajo Country about which we know nothing.

Rocks of this great antiquity are very difficult to study. In the first place, the original rock type—igneous, sedimentary, or another metamorphic—cannot always be determined. Secondly, they cannot usually be correlated from one exposure to the next. For example, metamorphic rocks in the San Juan Mountains look and date about the same as those in Grand Canyon, but it is not known whether they are continuous, having been formed as a single body by the same processes. A third problem is that the rock bodies have been so highly contorted that it usually cannot be determined by studying them in the field which side was originally up and which part of the rock body is younger than another. Consequently, the rocks must be age-dated by studying their radioactive minerals. The ratio of the remaining parent radioactive element, such as uranium, to the amount of daughter decay material, such as lead, is used to determine a calculated decay rate. This method requires very expensive equipment to enable such minute measurements, and the decay rates applied to find the solution are not completely proven and may be in error. To make matters worse, one must find rocks that contain radioactive minerals, and that may not be easy.

Rocks older than about 570 million years are said to have been formed in Precambrian time. If the Earth is about 4.5 billion years old, some 4 billion years of Earth history have been lumped into one time frame, and the rock record of much of that time is gone, or is indeterminable—truly lost to antiquity.

THE GREAT UNCONFORMITY

As mentioned above, the erosional surface at the top of the crystalline basement in Grand Canyon has been called The Great Unconformity. Except in isolated down-faulted pockets, rocks of Paleozoic age—namely the Tapeats Sandstone—rest directly on the metamorphic basement complex, the Vishnu Schist. I stated earlier that metamorphic rocks form under conditions of immense heat and pressure, requiring great depths of burial, but I have not yet explained where those great thicknesses of cover rocks go. Everywhere on our planet there is a sharp break between the metamorphic complex below, the older Precambrian basement, and sedimentary rocks that show little or no effects of high-grade metamorphism. The contact between the two distinctive rock types is invariably smooth and sharp; there are no gradational events recorded. In the case of the Grand Canyon exposures, there is a hiatus of about 1 billion years of Earth history that has left no record. What kind of rock buried the crystalline basement sufficiently deep to create an environment where intense metamorphism could occur? Certainly whatever it may have been was eroded away, but where did it go? Sedimentary rocks of the younger Precambrian Grand Canyon Supergroup must represent a second generation of deposition of the mysterious and elusive results of that erosion, but geologists do not know where the source of the thick red-bed section could have been, except that it was to the west of Grand Canyon. These questions without answers are some of the greatest unsolved mysteries in geology.

YOUNGER PRECAMBRIAN SEDIMENTARY ROCKS

Relatively unmetamorphosed sedimentary rocks, overlying the metamorphic basement and underlying rocks of Paleozoic age, are found in many places in the western United States. The Grand Canyon Supergroup, especially the excellent exposures in eastern Grand Canyon, is a classic example. There, nearly 13,000 feet of sedimentary rocks, mostly red beds, have been preserved from erosion in a giant down-faulted wedge on The Great Unconformity. The section is apparent from viewpoints on the rims in eastern Grand Canyon. Obviously, this great thickness of sediments was deposited across a wide expanse of the countryside, but only isolated patches have remained as local hints of another era where faulting saved bits of the section from intense planation at the erosional surface. Other areas where smaller exposures may be seen in Grand Canyon are near Phantom Ranch and at Bass Camp in the central canyon.

It should be noted here that the intense, high-temperature kneading that formed the Vishnu Schist was completed long before the Grand Canyon Supergroup was deposited, as there is precious little metamorphism evident in the sedimentary rock cover.

The Precambrian sedimentary section consists of a basal boulder bed of weathered metamorphic rocks known as the Hotauta Conglomerate, easily viewed at the foot of Hance Rapid on the Colorado River. The scattered patches of boulders are overlain by tidal-flat sedimentary deposits of the Bass Limestone. The Bass is commonly associated with a dark, igneous sill that intruded the carbonate rock in Precambrian time, forming an obvious white seam of asbestos at the baked contact. Above the Bass is an interval of reddish brown mudstones and siltstones of the slope-forming Hakatai Shale, locally punctuated by black igneous dikes. A thick section of apparently marine sandstone of the Shinumo Quartzite forms a cliff-ridden gorge stratigraphically above the Hakatai near Nevills Rapid, followed upward by several thousand feet of more sandstones, red shales and siltstones of the Dox Formation. After the Dox was laid down as coastal lowland deposits, an episode of volcanic activity resulted in basaltic lava flows now preserved as the Cardenas Lavas, which date at around 1 billion years old. All together, these formations comprise the Unkar Group. More sandstone of the Nankoweap Formation was deposited on the ancient lava flows, followed by a complex sequence of red beds and thin limestones of the Chuar Group, preserved only in a complexly faulted area north of the river in easternmost Grand Canyon.

Grand Canyon is the only place this section is exposed in or near Navajo Country, and only very rare wells have been drilled to sufficient depths to encounter the section elsewhere. Quartzites similar to the Shinumo are exposed in several places, however, indicating that sedimentary rocks were deposited over much of the American West prior to Paleozoic time. For example, great thicknesses—10,000–20,000 feet—of sedimentary rocks, mostly quartzites, occur in the Belt Supergroup in the Northern Rocky Mountains of western Montana and northern Idaho. Some 10,000 feet of mostly quartzite, with some purplish shale, comprises the core of the Uinta Mountains in northeastern Utah, and a similar section of perhaps 10,000 feet of quartzite and dark-colored slate (metamorphosed shale) form the Grenadier Range atop the San Juan Mountains of southwestern Colorado.

In Navajo Country proper, quartzites and slates nearly identical to the San Juan Mountains section occur in the Tusas and Truchas ranges north of Santa Fe and west of the Rio Grande Rift. These rocks are caught up in a wrench fault zone that may extend from the northwest corner to the southeast corner of the United States, crossing through the San Juan Mountains, the north-central New Mexico ranges, the Texas Panhandle, and southern Oklahoma. In the San Juan Mountains where the faults can be dated approximately, they must have been moving between 1.78 and 1.52 billion years ago.

In south-central Navajo Country, similar quartzites may be found near Fort Defiance in the high Defiance Plateau that lies just west of the Chuska Mountains. There red rocks of the Permian System (one can call them variously Abo Formation if New Mexico terminology is used, Supai Formation if Arizona names are used, or lower Cutler Formation if you come from Colorado) rest directly on the Precambrian quartzite. Thus, the Defiance Uplift must have been a high feature during all of Paleozoic time prior to the Permian.

In western Navajo Country, the exposures described in Grand Canyon prevail.

TIMES OF EARLY LIFE

FOUNDATION ROCKS

CAMBRIAN SYSTEM

The Paleozoic era dawned on Earth about 570 million years ago, but sedimentary rocks of Cambrian age were not deposited in Navajo Country until late in that geologic period. These were at first sandstones, deposited on sandy beaches at the eastern shoreline of a rising sea that shifted ever so slowly eastward onto the continental margin from the great Cordilleran trough, or seaway, that lay to the west of Navajo Country in what has become Nevada and western Utah. The encroaching shoreline and its beaches marched slowly eastward through Grand Canyon, eventually arriving in the Four Corners region and the San Juan Mountains in latest Cambrian time. For reasons not apparent, the shoreline skirted the heart of Navajo Country, leaving deposits of the Tapeats Sandstone in the northwest, and continued across the Monument Upwarp into southwestern Colorado, where it is known as the Ignacio Formation. It carefully avoided most of the Black Mesa and San Juan basins of northern Arizona and New Mexico.

Similarly, deposits of mudstones and siltstones containing fossils of marine organisms, such as trilobites and simple brachiopods, were deposited farther offshore, or west of the shoreline sands. These rocks, generally known as the Bright Angel Shale, skirt Navajo Country to the north and west, known to crop out only in eastern Grand and lower Marble canyons, and in the San Juan Mountains. The formation is present in the deep subsurface of the northern Black Mesa and San Juan basins, not extending southward into central Arizona or New Mexico. Also, tidal-flat lime mudstones of the Muav Limestone and an overlying unnamed dolomite that interfingers with and overlies the Bright Angel Shale, are seen in eastern Grand and lower Marble canyons, occurring in the subsurface only northward into east-central Utah (see figure 4), thus affecting the history of Navajo Country very little.

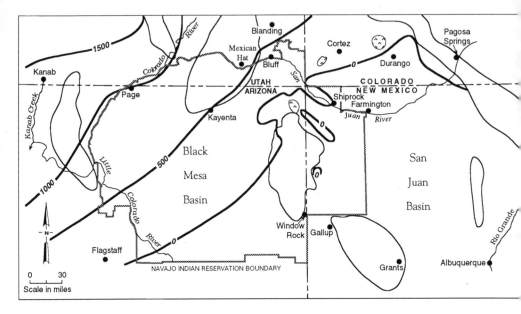

F I G U R E 4   Distribution and Thickness of Rocks of Cambrian Age in Navajo Country. Contours are lines connecting points of equal thickness (shown in feet) of these strata, indicating that rocks of Cambrian age thin from northwest to southeast across the region and are missing in northwestern New Mexico except in the immediate vicinity of the Four Corners area. Outlines of major geologic uplifts (see figure 2) are shown in light lines.

ORDOVICIAN AND SILURIAN SYSTEMS

Rocks of Ordovician and Silurian ages are not known to be present anywhere in the Colorado Plateau Province, and certainly not in Navajo Country. If sedimentary rocks of these two geologic periods were ever deposited here, they were cleanly and thoroughly removed by erosion from this region prior to Late Devonian time—another gap in our knowledge of geologic history and another great mystery.

DEVONIAN SYSTEM

Dry land prevailed across the Colorado Plateau and Southern Rocky Mountain provinces from Ordovician time through the Lower and Middle Devonian; no rocks of these ages are known to exist anywhere east of the Great Basin. Silty dolomites and dolomitic siltstones of Late Devonian age thin rapidly eastward through Grand Canyon, from a maximum thickness of about 1,200 feet in the western canyon to zero, with thin, local channel-fill deposits in lower Marble Canyon. There, the rocks are known as the Temple Butte Formation. In the easternmost

CHAPTER FIVE

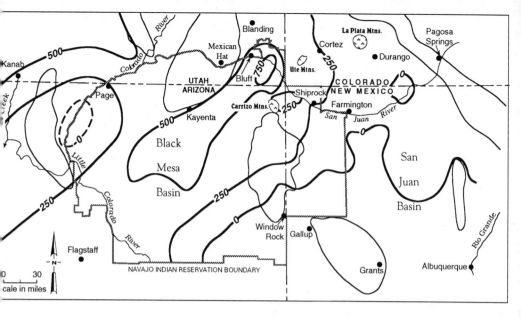

FIGURE 5   Distribution and Thickness of Rocks of Late Devonian Age in Navajo Country. Contours are lines connecting points of equal thickness (shown in feet) Sedimentary rocks of this age are missing throughout the southeastern part of the map area. Outlines of major geologic uplifts (see figure 2) are shown in light lines.

exposures in Marble Canyon, bedding features indicate that the sediments were deposited in channels eroded into the underlying Cambrian rocks as point bar deposits, much like in modern stream deposits. No fossils have been found in these channel-fill sediments, so it is not clear whether they were river or tidal channel sediments. At any rate, these channels represent very close proximity to the Late Devonian shoreline, at least in this region.

No holes sufficiently deep to encounter Devonian strata have been drilled from the Marble Platform eastward to near the Four Corners region. There, however, Late Devonian sedimentary rocks, known as the Aneth, Elbert and Ouray Formations, have been penetrated by drill holes as far west as the northwestern nose of the Defiance Uplift in northernmost Arizona, where Texaco discovered small amounts of oil in the McCracken Sandstone Member of the Elbert Formation in the late 1950s. These formations are not exposed on the crest of the Defiance Uplift, nor in the Zuni or Nacimiento mountains, indicating that these features were already high in Late Devonian time (see figure 5). These rocks are only exposed in the vicinity of Navajo Country in the San Juan Mountains of southwestern Colorado.

The Aneth Formation is only known to occur in a large "puddle" of marine deposits in the immediate Four Corners region. The dark-colored dolomite cannot be traced beyond these local occurrences in the deep subsurface, yet it is of definite Late Devonian age and of marine origin, as demonstrated by the fossil bony plates of primitive fish found in cores taken in the Shell Oil Company, Bluff No. 1 well, drilled on McCracken Mesa northwest of Bluff, Utah in 1954. A single possible thin exposure of the Aneth Formation has been tentatively identified high on the shoulder of Snowdon Peak in the San Juan Mountains.

The Elbert Formation, on the other hand, has been identified in deep wells from northernmost Arizona and New Mexico across the Colorado Plateau and Southern Rocky Mountain provinces. The base of the formation consists of the McCracken Sandstone Member, found locally in the subsurface of the Four Corners region, on ancient high fault blocks underlying some of the salt-intruded anticlines of the Paradox Basin, and exposed in the San Juan Mountains. The member produces oil and gas in the Lisbon Field south of Moab, Utah, and some gas has been produced from holes drilled on the Beautiful Mountain anticline of the east-central Chuska Mountains. The basal sandstone is thought to represent nearshore coastal sand deposits, while the unnamed upper member of the Elbert Formation has been identified as dolomitic tidal-flat deposits beneath most of the country from the Four Corners northward.

Intertidal deposits of the Elbert Formation change gradually upward into limestones of the Ouray Formation deposited in open marine conditions, as shown by fossils, consisting mainly of brachiopods, crinoids and the microscopic shells of single-celled Protozoa (class Foraminifera). Brachiopods are fairly primitive animals with two hinged shells, something like a clam; crinoids are "stemmed" animals that look much like flowers. Although it is not popular to have a single formation cross time boundaries, the fossils indicate that limestones of the Ouray Formation were first deposited in latest Devonian time, and deposition continued more or less continuously into earliest Mississippian time.

MISSISSIPPIAN SYSTEM

As Mississippian time unfolded, the sea slowly but surely began to withdraw from the Four Corners region westward, toward the Cordilleran seaway that lay in what is now western Utah and Nevada. In eastern Grand and lower Marble canyons, an erosional surface marks a break in sedimentation. In northern Navajo Country, however, the record of

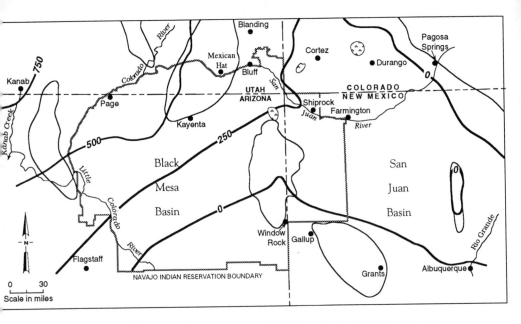

FIGURE 6 Distribution and Thickness of Rocks of Mississippian Age in Navajo Country. Contours are lines connecting points of equal thickness (shown in feet) indicating that rocks of this age are missing in exposures on the southern Defiance and Nacimiento uplifts and are completely missing on the Zuni Uplift. Outlines of major geologic uplifts (see figure 2) are shown in light lines.

the marine withdrawal is gradual and continuous. Dolomites of tidal-flat origin gradually replaced the Ouray marine limestones upward, dominating the massive carbonate rock unit called the Leadville Formation. This marine regression ended with exposure of the lower Leadville dolomites to weathering that caused a subtle disconformity to form near mid-section of the formation. This disconformity was recognized in Grand and Marble canyons by Eddie McKee and reported in his tome on the Redwall Limestone, and across the Colorado Plateau in deep wells and in San Juan Mountains exposures by me. The time of exposure of the rocks to weathering was only brief, as shallow but normal marine conditions returned across the entire region, still within Lower Mississippian time (see figure 6).

Above the regional disconformity, massive limestone beds of the upper Redwall and Leadville formations were deposited across the Colorado Plateau and Rocky Mountains. Widespread deposits of lime mud blanketed the provinces with local mounds, or banks, consisting of little more than crinoid fossil debris thickened over shallow shoals created by minor rejuvenation of the basement fault blocks. For some unknown reason(s), crinoids dominated the Mississippian sea floor and proliferated on shallow shoals to form large banks or mounds. Crinoids are marine animals that lived attached to the sea floor by holdfasts that resemble plant roots, stood well off the bottom on segmented stalks that resemble plant stems, and lived in calyces (heads) protected by calcite plates with protruding segmented arms for feeding. These critters looked much like flowering plants, but were actually animals of the phylum Echinodermata that disintegrated into piles of segmented calcite parts to become sedimentary grains that formed the limestones. They were especially dominant in shallow-water deposits. When dolomitized after deposition, the crinoidal segments were readily dissolved to form highly porous rocks that made excellent reservoirs for petroleum deposits. Several oil-producing layers of this sort have been found north of Navajo Country at Lisbon Valley, Big Flat and Salt Wash fields in eastern Utah.

The sea withdrew westward, and the limy deposits of the Leadville and Redwall formations were again exposed to severe weathering conditions, on this occasion for most of Late Mississippian time and perhaps some of the Early Pennsylvanian Period. Deep weathering of the limestone countryside produced thick lateritic soils in the Four Corners Region, known as the Molas Formation, with drainage flowing westward toward the seaway via deep estuarine channels recognized only recently as the Surprise Valley Formation in western Grand Canyon.

The stage was then set for one of most dramatic episodes of mountain building ever to be recognized in the geologic record since the formation of the Great Unconformity.

PENNSYLVANIAN SYSTEM

As the time we call the Pennsylvanian Period dawned, the countryside that is now Navajo Country was a flat lowland, simmering in moist, hot climatic conditions that resembled those of today's southeastern United States. The limestone terrain was cloaked with a thick cover of deep red soil, and drainage was largely accomplished via an underground network of naturally formed caves and tunnels. Sinkholes dotted the landscape. A thin layer of red mudstone that is nearly ubiquitous throughout the Colorado Plateau and Southern Rocky Mountains, the Molas Formation, is testimony to this scenario. Gradual changes in geological structural style slowly evolved on a global scale, laboring to reform the landscape for millions of years to come.

In North America we call this period the Ancestral Rockies orogeny

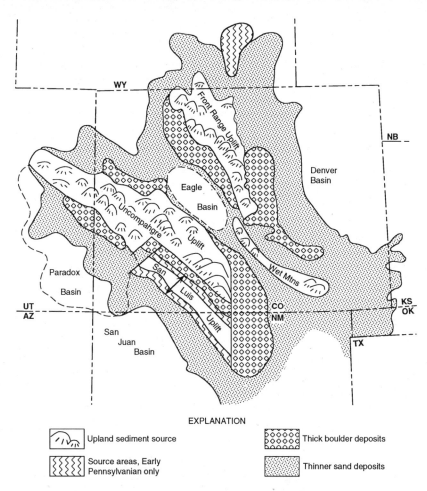

EXPLANATION

Upland sediment source	Thick boulder deposits
Source areas, Early Pennsylvanian only	Thinner sand deposits

FIGURE 7   The Ancestral Rocky Mountains of Pennsylvanian Age and Their Associated Basins of Sedimentary Deposition. Although these ancient mountain systems are not located in Navajo Country, sediments derived from the high source areas are extensive in rocks of Pennsylvanian and Permian age. This regional map has been modified from W.W. Mallory (1972, 132).

(an orogeny is an episode of mountain-building activity); in Europe and Asia it is called the Hercynian orogeny. Plate tectonics specialists attribute the widespread geologic unrest to collisions of the continental plates; whatever the underlying cause or causes, great compressional forces became hyperactive in the Earth's crust. Great mountain chains arose to unprecedented heights, and intervening faulted depressions controlled the distribution of seas across previously quiet, continental regions (see figure 7). Navajo Country, although not hosting the brunt of the structural onslaught, was affected deeply.

The region now known as Colorado was perhaps most involved in the structural disarray that followed. Fault blocks of magnificent proportions gradually emerged from the previously quiet landscape, and intervening down-faulted basins gradually filled with seawater. In Colorado, a mountain range arose at about the present-day site of the Front Range. Other uplands emerged where the Sawatch Range now reigns and the Wet Mountains lie less conspicuously in the central Southern Rocky Mountains, both separated from the Front Range uplift by the down-faulted Eagle Basin, also known as the Central Colorado Trough. Closest to Navajo Country, the Uncompahgre Uplift extended northwestward across southwestern Colorado into east-central Utah. All of the uplifts were attacked by the greatly intensified forces of erosion, shedding vast volumes of sand, gravel and boulders that would be distributed by alpine streams, only to accumulate in the adjacent structural basins. Thousands of feet of red sandstones and conglomerates line the margins of the basins immediately adjacent to the up-faulted mountain masses; any space left over in the basins was filled with evaporites (precipitated salt and gypsum) in the structurally restricted marine circulation within the partially enclosed basins.

The largest and perhaps most significant of these down-faulted basins, the Paradox Basin, lies to the immediate southwest of the faulted flank of the Uncompahgre Uplift just north of Navajo Country. There, red sandstones and conglomerates reach a maximum thickness of perhaps 18,000 feet adjacent to the uplift, and the remainder of the basin is filled with salt and related evaporites that reached depositional thicknesses exceeding 4,000 feet. The only definitely established entryway to the basin for marine waters was through the present-day San Juan Basin, although other possible, probably minor, sources of normal seawater may have fed salts episodically to the basin through the Black Mesa Basin and a narrow gap between the San Rafael Swell and the Uncompahgre Uplift in the vicinity of Price, Utah.

The climate was by now hot and arid, and the Paradox and Eagle basins acted as natural precipitation tanks for the minerals dissolved during the episodic seawater feeding periods. These influxes of marine water into the basins are considered sporadic because sedimentary rocks of this age indicate repeated advances and retreats of shorelines—the rise and fall of sea level on a worldwide basis. These global sea level fluctuations are attributed to glacial cycles in the polar regions, just as sea level fluctuated with repeated glacial episodes during the so-called "Ice Ages" of the past million or so years. Indeed, sedimentary rocks of the deep Paradox Basin and its flanking shallow shelves are cyclic in nature.

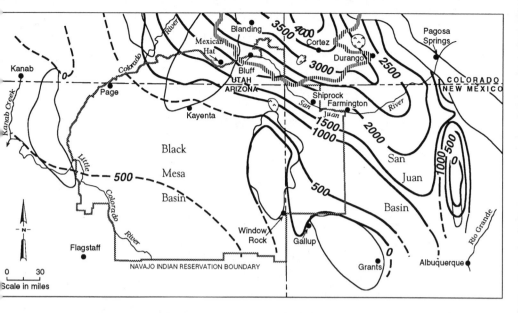

FIGURE 8   Distribution and Thickness of Rocks of Pennsylvanian Age in Navajo Country. Contours are lines connecting points of equal thickness (shown in feet) of strata of this age. A narrow seaway entered the otherwise restricted deep Paradox Basin of southwestern Colorado and southeastern Utah through the San Juan Basin in New Mexico; the patterned line represents the southern limits of salt deposition in the Paradox Basin. Dashed lines in Arizona show that very few wells have penetrated Pennsylvanian rocks in the Black Mesa Basin, and thicknesses there are conjectural. Outlines of major geologic uplifts (see figure 2) are shown in light lines.

Deposits of salt and other evaporites extend in the subsurface southward only into the immediate area of the Four Corners and northern Navajo Country, being limited near the northern Arizona and New Mexico boundaries. The Akah cycle is the most extensive of 29 salt cycles. Adjacent to the margins of each of the salt cycles, limestones—deposited cyclically along the southern shelf of the basin in shallow, open marine environments—dominate the Pennsylvanian System (see figure 8). These cycles of carbonate rocks host myriad buildups, or shoals, constructed by the proliferation of calcareous algae (simple, non-vascular, but calcified marine plants) under ideal environmental conditions of shallow water and consequent intense light where chemical nutrients are abundant. As the rocks in these mounds, or sedimentary buildups, are composed of chips of algal fragments, naturally stacked like corn flakes, they are locally highly porous and thus ideal as reservoir rocks for petroleum accumulation when found in the subsurface. Numerous examples of these porous limestone accumulations occur in the canyons of the San Juan River between Bluff and Mexican Hat, and again through the Goosenecks downstream from Mexican Hat. Because of the great economic significance of these petroleum reservoirs within northern Navajo Country, they will be discussed in more detail in later chapters on the San Juan River canyons and petroleum occurrences.

Somewhat surprisingly, Middle Pennsylvanian time is not represented by rocks in lower parts of the Supai Group in Grand and Marble canyons. There, only rocks of the early and late Pennsylvanian system have been preserved. Middle sections, if ever deposited, have been removed by erosion and are now represented only by a disconformity. It is apparent that regional uplift occurred during Middle Pennsylvanian time along the East Kaibab and Echo Cliffs monoclines.

Thus, much of Navajo Country is underlain by thick marine deposits of Pennsylvanian age, especially in the San Juan Basin and San Juan River canyon regions. However, they are only seen at the surface in the deep San Juan canyons and to the south of Navajo Country in uplifted high mountain ranges that border the Rio Grande and along the Mogollon Rim country. Uplifts within Navajo Country, the Zuni and Defiance (Chuska) Mountains, were forced upward as satellite positive elements of the Ancestral Rockies during this time period and did not receive sedimentary deposits of Pennsylvanian age (see figure 8). There, as well as in the high Nacimiento Mountains, specifically on San Pedro Mountain, rocks of the younger Permian System were deposited directly over Precambrian basement rocks, with strata of Cambrian through Pennsylvanian age lapping onto their flanks as seen by deep drilling.

Near the close of Pennsylvanian time, during what geologists recognize as the Virgilian Series, sea level lowered sufficiently that marine waters deserted the Four Corners region and most of Navajo Country, only to return to the outer fringes of this great land during Permian time.

CHAPTER SIX

RED BEDS INHERIT THE EARTH

PERMIAN SYSTEM

Although the sea had withdrawn from the Colorado Plateau and Navajo Country near the close of Pennsylvanian time, large volumes of sand, mud, gravel and boulders continued to overwhelm the lowlands, washing down from the adjacent Uncompahgre Uplift, through Early Permian time. The deep structural trough that was nearest the source area, the Uncompahgre uplift, trapped the coarser-grained sedimentary debris, but the finer mud and sand was spread far and wide across the Colorado Plateau by streams. Coarse-grained arkosic (containing feldspar) sandstones and conglomerates were trapped close to the uplifted source area to form stream deposits that attained thicknesses of several thousand feet, now known as the Cutler Formation. However, the finer-grained sediments formed coastal lowland and tidal-flat deposits that are readily discernable into distinctly separable formations. All of these deposits are reddish brown in color.

The country rock exposed to erosion on the Uncompahgre uplands consisted of metamorphic rocks that were high in feldspar content and contained abundant ferromagnesian minerals, such as hornblende, augite and chlorite. The abundant feldspar that was delivered to the lowlands by streams along with the dominant white quartz sand, gave a pinkish cast to the rocks. Ferromagnesian minerals, those that contain abundant iron, magnesium and silicates, were deposited along with the feldspar-rich sand, but were more susceptible to chemical weathering and decomposed to release their iron content into groundwaters. As the iron oxidized (rusted) in place, red and brown minerals such as hematite, goethite and limonite formed, as well as others, and were distributed throughout the sedimentary accumulation to stain sand grains or act as a red matrix between other grains, resulting eventually in reddish sandstones, mudstones and shales. Such iron-rich sediments deposited in arid lowland environments produced red rocks now found in profusion across the Colorado Plateau and Navajo Country in Permian time.

As the finer-grained sediments were distributed across the coastal lowlands that spread far afield west and south of the Uncompahgre source area, they separated mechanically according to the environments in which they were transported and deposited. Fine-grained sediments (mud and silt) were deposited in quiet environments where streams and tidal channels distributed mud and silt by sluggish and ponded flow. Deposition of red mud was occasionally interrupted by incursions of shallow marine sand accumulations and the development of desert dune fields that periodically covered vast regions of the province. Dune sands were sometimes blown from the sand-laden Cutler deposits, but often came from geographically separate sources.

As we have seen, the Permian System, or Period, was originally defined for rocks studied in the southern Ural Mountains of Russia and Kazakhstan. Rocks of similar age and stratigraphic position throughout the world must be compared directly with the Russian section by study-ing the similarity of fossil evolutionary trends to derive relative ages. Until recently, there was much confusion as to where to draw the boundaries of the Permian System, due to disagreements among Russian geologists. In 1991, however, they reached a consensus at the Interna-tional Congress on the Permian System of the World, held in Perm, Rus-sia, that the base of the Permian System should be placed at a particular bed in the stratigraphic succession where important changes occur in certain fossil groups.

Although many kinds of fossils are present in the rocks and are sometimes useful for correlation, namely brachiopods, cephalopods, conodonts, plant spores and pollen, a single variety of fossils known as fusulinids have been widely used to establish global age correlations in rocks of late Paleozoic age. Fusulinids are small, generally 0.25–0.5 inches in length, football-shaped animals of the phylum Protozoa, class Foraminifera. Although single celled, the fossils consist of layers of chambered, elongated sheets of tubes, curled in jelly-roll fashion to form very complex internal structures. To study these complicated little critters, thin sections (slices of rock cut sufficiently thin to permit light to pass through) must be cut exactly through the initial, juvenile cham-ber, and measurements of all internal structures must be made with the use of microscopes.

Fusulinids evolved very rapidly through the Pennsylvanian and Per-mian periods, making them excellent tools for worldwide correlations. Pennsylvanian fusulinids are generally elongate, or fusiform, and increase in size as they occur higher in the section in younger rocks. For some unknown reason, they suddenly became greatly inflated, more basketball-shaped, at a point in time that is coincident with the chosen bed defined as the base of the Permian System by Russian paleontologists. The base of the Permian System can thus be determined in North America by studying these complex, but handy little fossils.

In the Ural Mountains the basal beds now assigned to the Permian System are called the Asselian Stage. Rocks of the same relative age in North America have long been assigned to the Wolfcampian Series, named for sedimentary rocks found in the Wolfcamp Hills of West Texas. Because of an error in judgement by American paleontologists studying cephalopods, the base of the Permian was placed lower in the section, but that was for decades considered to be acceptable because of the confused state of Russian stratigraphy. Since the 1991 consensus in Russia, the base of the Permian System has been raised to the original base of the Wolfcampian beds, where inflated fusulinids first occur. This development will result in some departure in age assignments used here from previous usage in older geological publications.

HALGAITO SHALE

The sea withdrew from the Colorado Plateau region in Late Pennsylvanian time, and a surface of widespread weathering and erosion—an unconformity—formed across the province. The next recorded event was the formation of a marine embayment that entered the northwestern corner of the former Paradox Basin as an extension of the large Oquirrh (pronounced oh!-kur) Basin in central Utah. Rocks deposited in this broad embayment, known as the Elephant Canyon Formation, are thickest and best exposed near the confluence of the Green and Colorado rivers in Canyonlands National Park. They extend onto the northern Monument Upwarp, but do not occur in Navajo Country.

The significance of these rocks to our story is that limestones of the Elephant Canyon Formation, datable by studying their contained fusulinids, interfinger southward with the earliest red-bed formation, the Halgaito Shale (pronounced Hal-khigh-toh!), and thus that continental lowland and tidal-flat deposit can be dated rather closely. Rocks and fossils of the Elephant Canyon Formation, and therefore the Halgaito Shale, were originally believed to be of earliest Permian age. Recent changes in the stratigraphic position of the Pennsylvanian-Permian boundary, however, now place these rocks in the uppermost Pennsylvanian System (the Virgilian Series), with only the uppermost beds of the Elephant Canyon Formation containing truly Lower Permian fusulinids.

Red rocks beautifully displayed in the lower slopes beneath Cedar Mesa north of the San Juan River on the Monument Upwarp, and in the spectacular display of the syncline (downfold) in and near the Valley of the Gods just north of Mexican Hat are exposures of the Halgaito Shale. The formation name is Navajo, meaning "a spring in an open valley"; the type section is in the southern reaches of the Mexican Hat syncline south of the San Juan River. Previously referred to as of earliest Permian age, the Halgaito must now be considered to be of latest Pennsylvanian (Virgilian) age. The red-bed formation was deposited on a broad coastal lowland and/or intertidal flats, the two closely related depositional environments being almost indistinguishable in the rock record. Meandering stream or tidal channels found in the formation near Mexican Hat contain fragments of vertebrate bones of amphibians and primitive reptiles.

CEDAR MESA SANDSTONE

Great volumes of relatively clean quartz sand flooded the central Colorado Plateau, burying the marine deposits of the Elephant Canyon Formation and the intimately related tidal flats and continental lowlands represented by the Halgaito Shale. These resulting sandstones are known as the Cedar Mesa Sandstone for prominent exposures that form the cliffy margins of Cedar Mesa north of Mexican Hat, Utah. The source of the sand has been attributed by some geologists to be the Uncompahgre Uplift. However, the lack of feldspar and ferromagnesian minerals make that source seem unlikely. Cross-bedding studies clearly indicate that the sand was deposited by currents moving from the northwest, making a separate northwesterly source area an appealing option. Wherever the sand originated, it clearly interfingers with limestones and shales of the uppermost Elephant Canyon Formation, indicating that the incursion of sand into the region began in Early Permian time as now recognized.

It has been shown that cyclic fluctuations in sea level were the trademark of rocks of Pennsylvanian age throughout the world. Characteristics of the Elephant Canyon Formation and the related Cedar Mesa Sandstone clearly indicate that cyclic deposition continued into early Permian time, although quartz sand became the dominant sediment type. The Cedar Mesa consists of white, highly cross-bedded quartz sand that occurs in thick individual beds separated by nearly flat bedding surfaces. The nature of the overall bedding suggests that the sand was deposited in a coastal environment. An oscillating sea level first encroached onto the coastal lowlands, allowing nearshore water-laid sand bars to be deposited, only to retreat later, encouraging vegetation and groundwater-controlled compaction to fix the former sand deposits and permit episodes of coastal windblown dunes to dominate the abandoned lowlands. The process was repeated many times to permit the accumulation of up to 1,200 feet of Cedar Mesa Sandstone, the thickest deposits being found just to the northwest of Navajo Country.

The massive body of the Cedar Mesa Sandstone crosses the northwestern corner of Navajo Country in the subsurface from the ominous cliff-forming exposures bordering Cedar Mesa, and emerges in Marble and Grand canyons as the Esplanade Sandstone, the uppermost formation of the Supai Group. The eastern limit of the huge sand body is fascinating, as the clean white sandstones of the Cedar Mesa change aspect in very short distances, interfingering with red, arkosic sandstones of the Cutler Formation in the Needles District of Canyonlands National Park. South of the Abajo Mountains, the change in character is more abrupt, with white sandstone changing to red siltstones and bedded gypsum in distances of only a quarter-mile, the band of abrupt change occurring along the high side of the Comb Ridge monocline near the crest of the Lime Ridge anticline. The abrupt change in depositional character suggests that the eastern margin of the Monument Upwarp was sufficiently high to impede circulation and that lagoonal or highly restricted intertidal environments prevailed east of the major structure.

Navajo Country is thus underlain by the gypsiferous red-bed facies of the Cedar Mesa east of the Monument Upwarp across the Four Corners region, and to the south in Black Mesa Basin. The red-bed facies of the formation is believed to be present on the Defiance Uplift west of Fort Defiance, where it provides the foundation for the Blue Canyon Dam and is meagerly represented by a couple of thin limestone beds, as seen near Hunters Point, south of Window Rock. The Defiance Uplift remained high in Cedar Mesa time, with these lower Permian strata lapping onto its flanks.

ORGAN ROCK SHALE

Coastal lowland and/or intertidal environments returned rather abruptly to the entire Colorado Plateau and Navajo Country following the cessation of Cedar Mesa sedimentation, and red fine-grained mud and silt again dominated the landscape. The resulting formation is known as the Organ Rock Shale in the Monument Valley region, where it forms the lower red slopes of the myriad buttes and spires. Geologists who have worked mostly in Arizona tend to call this unit the Supai Formation on the Defiance Uplift. However, it correlates directly with the Hermit Shale of Grand Canyon, the red slope-forming formation that overlies the Supai. There is little doubt when tracing the underlying sandstone from Canyonlands southwestward into Grand Canyon using deep drill holes, that the Cedar Mesa Sandstone is indeed the same rock unit as the Esplanade; the overlying red beds of the Organ Rock can be as easily correlated with the Hermit Shale.

East of the Four Corners area and the Defiance Uplift, in the general vicinity of the San Juan Basin and Nacimiento Mountains, the entire combined sequence of the Halgaito, Cedar Mesa and Organ Rock formations merges into one thick body of red beds. In Colorado the section is known as the Cutler Formation; in New Mexico it is called the Abo Formation. In either case, the red beds occur above rocks of Pennsylvanian age in the basins and lie directly on Precambrian basement rocks on the Defiance, Zuni and northern Nacimiento uplifts, all ancient highlands that persisted from Precambrian through Pennsylvanian times. The source of the red beds was unquestionably the Uncompahgre Uplift, for coarse arkosic sandstones grade perceptibly southward and westward into the finer-grained red shales, mudstones and siltstones of the Abo Formation, and westward into the Cedar Mesa gypsiferous red beds and the overlying Organ Rock Shale.

DECHELLY SANDSTONE

Sandstones, salmon red in color, comprise a prominent component of the Permian System in Navajo Country above the underlying red beds of the Abo–Organ Rock formations. This most scenic cliff-forming sandstone is the DeChelly (pronounced de-shay!) Sandstone, found in Monument Valley, where it forms the massive cliffs of the magical buttes and spires, and in the precipitous walls of Canyon de Chelly, from which its name is derived. New Mexico geologists, having worked toward the Four Corners region from a different perspective, call the same rocks the Meseta Blanca Sandstone Member of the Yeso Formation. The ubiquitous odd shade of red of the formation results from iron-oxide coatings, or stain, on each individual quartz sand grain, rather than from clay matrix within the sandstone.

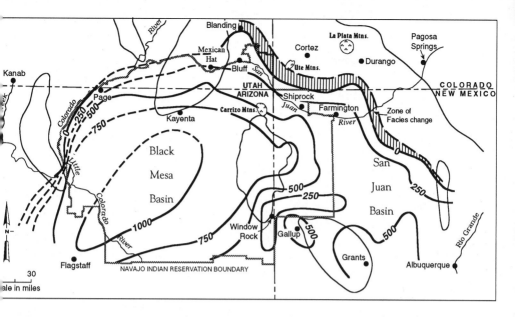

FIGURE 9 Distribution and Thickness of the DeChelly Sandstone, the Most Prominent Formation of Permian age in Navajo Country. Contours are lines that connect points of equal thickness (shown in feet) of eolian sandstone of Permian age. The orange sandstone is thickest in the Black Mesa Basin, but occurs widespread across the San Juan Basin, where it is called the Meseta Blanca Member of the Yeso Formation. It thins and grades northward into red beds of the Cutler Formation along the trend marked by the ruled pattern. Outlines of major geologic uplifts (see figure 2) are shown in light lines.

Regional studies show that the highly cross-bedded, salmon colored sandstone is one and the same, whatever name is applied, having been blown southward by winds crossing the Cutler sand flats from the north. Dune sands were deposited in the San Juan Basin in New Mexico and the Black Mesa Basin in Arizona, barely crossing the intervening Defiance Uplift where only thin deposits are found at Hunters Point, south of Window Rock, and Bonito Canyon, west of Fort Defiance. The sand body is thickest in Black Mesa Basin, thinning eastward to 825 feet in Canyon de Chelly. It thins further to less than 200 feet on the crest of the Defiance Uplift, but thickens again into the San Juan Basin (see figure 9). The DeChelly Sandstone is the uppermost formation of the Permian System in the Four Corners area.

Equivalent deposits of thick sandstones, named the Schnebly Hill Formation, occur along the Mogollon Rim, but they have the general appearance of having been waterlaid. The Schnebly Hill is largely equivalent in age and extent with the DeChelly Sandstone. Lying between the Four Corners region and the Mogollon Rim, age-equivalent rocks in the subsurface within the intervening Holbrook Basin change abruptly to mudstone, with interbedded gypsum and salt deposits. It is apparent that the Holbrook Basin was depressed structurally as the DeChelly–Schnebly Hill formations were being deposited, forming a local evaporite basin.

YESO FORMATION

The sea returned—at least to the southern margin of Navajo Country—yet again in Permian time. Rocks of marine origin, namely reddish mudstones and dolomite beds called the Yeso Formation, thin northward from central New Mexico into the region of the Zuni and southern Nacimiento mountains. There, the southeastern extension of the DeChelly Sandstone is called the Meseta Blanca Member of the Yeso; the upper marine portion of the Yeso is known as the San Ysidro Member. Red beds with thin dolomites characterize the formation along the southeastern margin of Navajo Country but grade abruptly northward into red mudstones indistinguishable from the Abo Formation, apparently marking a northern shoreline for the deposits along the southern flank of the San Juan Basin.

The upper member, seen in the red-bed facies near its northerly pinchout, is believed to occur on the southern Defiance Uplift at Hunters Point, overlying beds obviously belonging to the DeChelly Sandstone. There, thin but typical exposures of the DeChelly Sandstone are capped by slope-forming reddish mudstone beds, separating the DeChelly from an overlying light-colored sandstone typical of the younger Glorieta Sandstone to the east. If this relationship is correct, the Hunters Point exposure heals an otherwise dangling correlation between Arizona and New Mexico stratigraphy.

COCONINO-GLORIETA SANDSTONE

As if that weren't enough, sand then piled onto the southern fringes of the Colorado Plateau and Navajo Country. However, this time the source area was to the southwest; a highland, known as the Mazatzal Uplift arose in central Arizona, and copious amounts of quartz sand again became readily available.

The sand pile, known as the Coconino Sandstone, accumulated along the Mogollon Rim and northward into Grand Canyon, thinning to a pinchout approximately along the Arizona-Utah border. The white, highly cross-bedded sandstone thickens southward to more than 1,000 feet near Pine, Arizona, and grain size increases to the south as well. These features strongly suggest a southerly source area for the formation. Cross-bedding studies, however, contradict that option, as lee slopes on the petrified eolian dunes are strongly oriented in a southeasterly direction, implying that the sand must have been derived from the northwest. A plausible explanation would be that sand derived from the Mazatzal upland was carried northward by streams, only to be redistributed toward the south and southeast by prevailing winds. Whatever is the correct solution to the problem, the Coconino Sandstone is a light-colored, prominent cliff-forming unit in Grand and Marble canyons.

The distribution of the Coconino Sandstone is that of a giant fan, thinning to the west, north and east. The clean sand body appears to extend eastward into central New Mexico, where it is known as the Glorieta Sandstone. Although an apparent continuation of the Coconino sand body, the Glorieta takes on new character when it crosses the southern extension of the Defiance Uplift and the state line. It forms ledgy sandstone cliffs above the upper Yeso notch at Hunters Point, and rests directly on the true DeChelly Sandstone in Bonito Canyon west of Fort Defiance. From there on the crest of the Defiance Uplift eastward into New Mexico, the Glorieta is a thinner, rather uniformly distributed sandstone with smaller and more erratic cross-bedding sets, indicating that the sand was deposited probably in a shallow marine environment. The formation thins and pinches out a short distance north of the Zuni Mountains in the subsurface of the southern San Juan Basin, the apparent shoreline. Along the road between Fort Wingate and McGaffey in the western Zunis, large channels filled with Glorieta Sandstone cut through the underlying Yeso Formation, allowing the Glorieta to rest directly on Abo red beds. The Glorieta is seen only for a short distance northward in the southern Nacimiento Mountains, but is a prominent cliff-forming sandstone in central New Mexico.

KAIBAB-TOROWEAP-SAN ANDRES FORMATIONS

Finally, the last of the Paleozoic marine incursions took place, still restricted to the southern and western margins of Navajo Country. At first the lower unit, the Toroweap Formation, was mapped together with an upper, largely limestone unit, the Kaibab Limestone, in Grand and Marble canyons. However, Eddie McKee, in his classic studies in

Grand Canyon, recognized that the lower limestone and shale section was separated from the upper, mainly limestone unit by an unconformity that represents a withdrawal of the sea and sufficient justification for separation of the two units. The Toroweap Formation is a series of interbedded mudstone, limestone and dolomite beds, with gypsiferous deposits to the west. It changes in a very short distance to windblown sandstone, indistinguishable from the Coconino Sandstone, in its eastern limits near the mouth of the Little Colorado River and along the Mogollon Rim near Walnut Canyon, Arizona. The overlying limestone, the Kaibab Formation, forms the rimrock of Grand and Marble canyons, and floors the surfaces of vast plateaus in western Navajo Country.

The Kaibab extends as a plateau-supporting layer eastward across the Holbrook Basin and beneath the Petrified Forest but changes identity when it reaches New Mexico, where it is called the San Andres Formation. This direct correlation has long been recognized, but the two names are respected by state-line geologists. The San Andres is well exposed in the western Zuni Mountains, where limestone banks (bioherms) are to be found between Fort Wingate and McGaffey. To the north, in the subsurface of the San Juan Basin, the San Andres pinches out rapidly, but it thickens southward and locally contains evaporites south of the Zuni Mountains, again suggesting a slightly positive structural position for the Zuni highland during deposition of the San Andres.

The Kaibab–San Andres Formation cannot be traced northward past the southernmost fringes of Navajo Country but marks the close of sedimentation during Permian time on the Colorado Plateau. Thus, one of the most complex episodes in the geologic history of Navajo Country was sealed for posterity.

CHAPTER SEVEN

TRIASSIC RED BEDS

T here are no known rocks on the Colorado Plateau or Navajo Country dated as Late Permian in age. Half of Permian time is represented by non-deposition. If any rocks were ever deposited, they have been removed by pre-Triassic erosion. Thus rocks of Early Triassic age rest directly on mid-Permian or older deposits.

A marine seaway was present in the Great Basin country of modern-day western Utah and Nevada in Early Triassic time. The eastern effective shoreline vacillated somewhat over time, but generally it lay to the west of Grand Canyon and the San Rafael Swell. To the east, especially in Navajo Country, broad mud banks developed where fine-grained sediments were strewn across vast tidal flats that bordered the sea. The resulting chocolate brown mudstones and siltstones are called the Moenkopi Formation, named for exposures in Moenkopi Wash near the villages of Moenkopi and Tuba City, Arizona.

MOENKOPI FORMATION

Distribution of the Moenkopi Formation in Navajo Country is largely restricted to northeastern Arizona and eastern Utah, and it thickens generally westward to more than 300 feet as it interfingers with marine deposits. West of Grand Canyon, marine rocks of equivalent age reach thicknesses exceeding 2,000 feet. The Moenkopi Formation is made obvious by its characteristic dark reddish brown color and soft weathering features above light-colored, more resistant rocks of the Kaibab Formation in eastern Grand Canyon and southward across the southern Painted Desert region, and above the reddish sandstone cliffs of the DeChelly Sandstone in the vicinity of Monument Valley. The Moenkopi Formation thins to a pinchout eastward, and is not seen east of the Monument Upwarp nor on the Defiance Uplift north of Black Canyon

The Moenkopi Formation was divided into three members by Eddie McKee (1954). The lowermost, the Wupatki Member, consists of alter-

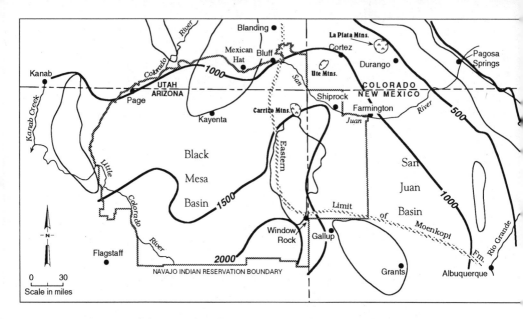

FIGURE 10 Rocks of Triassic Age. They are widespread across Navajo Country, as shown on this thickness (isopach) map (contours labelled in feet). The system is represented by the Moenkopi Formation that extends eastward to the stippled line, overlain by the Chinle Formation. Outlines of major geologic uplifts (see figure 2) are shown in light lines.

nating cliff-forming sandstone, thin bedded and ledgy siltstone, and crumbly, slope-forming mudstones—all reddish brown. The middle unit, named the Moqui Member, is generally made up of light-colored mudstones and siltstones that locally contain gypsum beds and nodules. Reddish brown alternating cliffs and slopes of fine-grained sandstone and siltstone with mudstone interbeds form the upper Holbrook Member, named earlier by Hager (1924). Mud ball conglomerates containing limestone clasts are common in the Holbrook Member.

The lower two members were deposited on extensive intertidal mud flats, as readily displayed by their classic sedimentary structures. Desiccation cracks (mud cracks) are prominent on many bedding planes, as are small, symmetrically shaped ripple marks that commonly occur in interference patterns (overlapping ripple patterns in differing trend orientations). Filled wandering tubes of burrowing organisms, such as worms and shrimp, are common, as are raindrop impressions. All signs lead to an interpretation of sedimentation on alternating wetted and dried surfaces of an intertidal environment of deposition for the Moenkopi red beds. The age assignment of Early to early-Middle Triassic is based largely on marine fossils, mainly ammonitic cephalopods, in closely related marine rocks that are found to the west.

The upper Holbrook Member is comprised of coarser-grained sediments and typified by the widespread occurrence of complex medium- to large-scale cross-beds that resemble sedimentary structures in modern stream deposits. Reptile tracks are commonly found on bedding surfaces, and fragmentary skeletal parts of reptiles, amphibians, and fish are commonly found in the sandstone and conglomerate interbeds. Following initial reports by Gregory (1917) of vertebrate remains near Holbrook and north of Winslow, Arizona, paleontologists have uncovered numerous bone fragments and teeth—even skulls—of primitive reptiles and amphibians in various exposures of the Holbrook Member. Scales and spines of fossil fish occur as far north as the Bears Ears on the Monument Upwarp north of Monument Valley. Most of the vertebrate remains have been found in sedimentary layers believed to have resulted from the accumulation of debris marginal to relatively fast-moving streams in point bar deposits, where animals and bone fragments carried by the streams naturally accumulate. Several quarries were opened near the towns of Holbrook, Winslow, and Cameron, and at Meteor Crater by a paleontological party from the University of California led by S.P. Welles in the late 1930s and 1940s, and considerable vertebrate fossil material was collected and cataloged. Of these collections, the amphibian remains are more numerous and better preserved than those of reptiles. However, reptilian tracks and trails are the most numerous, occurring mainly on ancient sand spits and mud flats on bedding planes in siltstone and sandstone. These were studied intensely by Frank Peabody, in conjunction with the University of California excavations, who recorded vertebrate trackways from most exposures he visited in northeastern Arizona. Fossil plant remains in the Holbrook Member are rare and poorly preserved. Occurring in fluvial deposits of the uppermost Holbrook Member of the Moenkopi Formation, above Lower Triassic fossiliferous equivalents of the lower two members, the Holbrook fauna is probably of Middle Triassic age.

No rocks recognized as Moenkopi are found on the Defiance Uplift north of about Hunters Point. Conglomeratic sandstones of the Shinarump Member of the Chinle Formation are present in deep channels cut into the Permian DeChelly Sandstone at Canyon de Chelly, at Nazlini Canyon, and atop the Defiance Plateau near the Chuska Mountains. This makes it obvious that the Defiance Uplift was high and being eroded by streams during Moenkopi time. However, reddish brown mudstones and sandstones, thought to have been deposited by streams, are present south and east of the Defiance Uplift on the flanks of the Zuni Uplift and the southern San Juan Basin, extending eastward to the eastern New Mexico plains (Lucas and Hayden 1989). These red beds have generally been called the Moenkopi Formation and are considered to be equivalent to, or slightly younger than, the upper Moenkopi to the west.

General conclusions regarding the Moenkopi Formation are that an open seaway lay to the west of Navajo Country in western Utah and Nevada in early Triassic time. Moenkopi sediments were deposited along the margin of the sea in vast tidal flats, occasionally interrupted by associated stream-dominated coastal lowland deposits, which extended as far east as the Defiance Uplift. As sea level stabilized and began to fall at the close of Moenkopi time (Middle Triassic), streams coursed the lowlands, distributing fluvial sediments in the wake of shoreline shifts back toward the west and in time covered the tidal-flat deposits of the retreating shoreline. Numerous varieties of vertebrate animals, including reptiles, amphibians, and fish, inhabited these coastal lowlands and streams.

CHINLE FORMATION

After the Moenkopi Formation was deposited, an interval of time followed when exposure of the lands and rather intense erosion, especially on the uplifted areas, occurred. Late Triassic sediments of continental origin, known as the Chinle Formation for rocks exposed near Chinle, Arizona at the mouth of Canyon de Chelly, blanketed Navajo Country. The initial deposits are found mostly in channels cut into the older underlying rocks to depths of 5–75 feet, and consist of stream-deposited coarse, gray sandstones and conglomerates of the Shinarump Member of the Chinle Formation. The Shinarump is a complex member that consists of laterally juxtaposed channel-fill deposits that may occur as scattered channels or as sheets of interlocking channel-fill deposits. The erosional surface it fills, however, clearly distinguishes the base of the Chinle Formation. The Shinarump Member occurs across most of Navajo Country, but it is not known to be present to the north in Canyonlands.

Younger geologists mispronounce this name as "Shin!-er-ump." The name, however, is derived from the Shinarump Cliffs near Kanab, Utah, which were named from the Paiute word for wolf, *Shinar*, (pronounced "Shin-air!") and "rump" for his rear end; thus, Shinarump (properly pronounced "Shin-air!-rump") is a more respectable way to say "Wolf-Rump Cliffs." Only mature geologists with even older professors from the Wild American West pronounce this correctly, thus displaying their true vintage and relative prowess. Major John Wesley Powell originally applied the name to these rocks near Kanab as early as 1873, but official designation of the conglomeratic sandstone was attributed to G.K. Gilbert in 1875.

The Chinle Formation, above the basal Shinarump Member, consists of varicolored mudstones and siltstones, with minor thin beds of sandstone and limestone stringers. Colors range from light gray to varied light shades of pink, red, green, brown and purple, often interspersed, mottled, and sometimes disrespectably garbled. It is no wonder that broad areas of exposure of the Chinle Formation are known as the Painted Desert. The formation was deposited in lakes and associated river flood plains on a broad, flat continental lowland environment.

The Petrified Forest Member is as much as 900 feet thick in Arizona, and is overlain by 150–400 feet of pink and red shales, mottled light green, with ledgy slopes of cherty limestone and sandstone of the Owl Rock Member, named for Owl Rock immediately west of Agathla Peak near Kayenta. In southeastern Utah, the Owl Rock Member is overlain by the reddish orange Church Rock Member, known in eastern Arizona as the Rock Point Member. To further complicate the named units within the Chinle Formation, sandstone beds such as the Hite bed and Lupton bed, occur locally within the section. The Chinle Formation is a formidable stratigraphic unit, as it attains a total thickness in excess of 1,500 feet along the southern reaches of Navajo Country in the Holbrook and southern San Juan basins (see figure 10).

For general purposes, most of us will be happy and psychologically fulfilled to ignore all of the variously named members and refer to the whole as the Chinle Formation: Rocks of the Painted Desert.

FOSSIL FORESTS

Petrified wood is a common constituent, especially in the lower part of the Chinle Formation, known as the Petrified Forest Member, but that doesn't mean that visitors to the Petrified Forest National Park or the Navajo Reservation should take any. Herbert Gregory, in his descriptions of the geology of the Navajo Country published by the United States Geological Survey (USGS) in 1917, was greatly impressed by the number, size, color and widespread distribution of petrified logs and stumps present in rocks of the Chinle Formation. He reported:

The abundance of fossil wood . . . is almost incredible, and its presence has made a profound impression on the native tribes. To the Navajos the logs are yeitsobitsin, the bones of yeitso, a monster who was destroyed by the sun and whose blood was congealed in lava flows. In the Piute mythology the broken trunks are the spent weapons of Shinarav, the great Wolf God; the accumulated masses mark the sites of battle fields. (Gregory 1917, 49)

Major Powell, who first used the term Shinarump Conglomerate for the basal Chinle member, may have had the Paiute mythological god in mind. Gregory goes on to say the following: "The tree trunks are very unevenly distributed. They usually occur in widely spread groups of unassorted large and small trees, all lying flat and trending in parallel or diverse directions, or overlying one another, like fallen timber in the path of a tornado." That extract paints a vivid picture for those of us who have lived in the Midwest.

Gregory described the occurrence of thousands of petrified logs in various localities across Navajo Country that measured from 30–80 feet in length, the largest having been found in Beautiful Valley, about mid-way between Chinle and Ganado, at 150 feet long and 4 feet in diameter. "Some of the logs are colored in harmony with the gray sands in which they are embedded, but most of them are colored by iron and manganese and assume beautiful tones of red, brown, yellow, and blue. Superoxidation has added brilliancy to colors on the surface of broken blocks, making the varicolored jasper, a much prized semiprecious gem stone." In places the fossil wood is found as carbonized fragments, and in eastern Monument Valley huge logs were found by miners that had been completely and delicately replaced by the canary yellow mineral carnotite, a rich uranium ore.

As to their occurrence, Gregory wrote as follows: "That the trees grew in the spots where they are now found is highly improbable. . . . It is believed that the tree trunks now turned to stone were carried by streams during floods. . . . The accumulations of trunks in the fossil forests are closely similar to piles of driftwood now seen along Colorado, Little Colorado, and San Juan rivers. . . ."

Most of the petrified wood was formed by the infilling of the cellular microfabric of the vascular deciduous trees, mostly pines, by microcrystalline silica. The mineral silica (SiO_2) is soluble in alkaline water and is thus readily transported and distributed in groundwaters. When acidic conditions are encountered, such as may be found in the cellulose wood fibers of dead and buried plant material, it is prone to precipitation. Thus, the buried wood is turned to a stonelike consistency, although the original cellular fabric may still be present as seen when petrified wood is examined under magnification, say 10X, with a hand lens. The wood has not been replaced, but the microstructure merely infilled, thus preserving the original wood texture and shape in detail. Coloration of the petrified wood is the result of the contaminants present in the silica; iron oxides produce red or brown, copper oxides form greens and blues, etc.

Today, much of the beautifully colored petrified wood of Navajo Country has been scavenged by rockhounds and other users of semiprecious stones. When many visitors each take a small souvenir, the treasure soon disappears forever, scattered to the four winds in meaningless little rocks that are soon discarded. Only in Petrified Forest National Park are the concentrations of large logs, such as described by Gregory, to be found in abundance and intact. It is little wonder that Petrified Forest Park Rangers are so jealously protective of the heritage left by the idiosyncracies of nature.

ENTER SAND

THE JURASSIC SYSTEM

GLEN CANYON GROUP

Layered rocks above the Chinle Formation, consisting almost entirely of sandstone—literally thousands of cubic miles of sandstone—are combined to form the Glen Canyon Group. The name was derived from magnificent exposures of the usually reddish colored sandstones in Glen Canyon of the Colorado River, now largely obscured by Lake Powell behind the Glen Canyon Dam. The canyon forms the northwestern boundary of the Navajo Indian Reservation between the mouth of the San Juan River and Lees Ferry, Arizona, where Marble Canyon begins. The group consists, in ascending order, of the Wingate Sandstone, the Moenave Sandstone, the Kayenta Formation and the Navajo Sandstone.

The type localities of all four formations lie within Navajo Country. The complex interfingering relationships of the different sandstone bodies were realized and published by Harshbarger, Repenning and Irwin in USGS Professional Paper 291 in 1957. Previous workers who studied the region for the USGS, especially Herbert Gregory and later Arthur Baker and his field associates, recognized and distinguished the stratigraphic units, but they also realized that there were highly complex, lateral and vertical relationships among the sandstone formations that remained to be resolved. It was not until the work of George Pipiringos and Bob O'Sullivan of the USGS, published in 1978, that the Jurassic section was found to be subdivided in the field by unconformities, breaks in the stratigraphic record that indicate episodes of erosion or non-deposition. These are designated the J-0 to J-5 unconformities, in ascending order. J-0 is the unconformity found at the top of the Chinle Formation throughout much of the Colorado Plateau. Recognition of these unconformable surfaces helps to bring order out of chaos in this complex system of great sand seas of the Jurassic Period.

Wingate Sandstone

The lower and older of the four formations of the Glen Canyon Group, the Wingate Sandstone, was named for the majestic red sandstone cliffs north of Fort Wingate, east of Gallup, New Mexico, by Dutton in 1885. Harshbarger and associates (1957) describe the formation thus:

> In the Navajo country, the Wingate sandstone comprises two mappable units. The lower unit consists of reddish-orange parallel-bedded, thin-bedded siltstone and sandstone. This unit is described as the Rock Point member from typical exposures near Rock Point School, Apache County, Ariz. The upper unit consists of reddish-brown, fine-grained sandstone which is crossbedded on a large scale and commonly forms vertical massive cliffs. This unit is herein referred to as the Lukachukai member, from typical exposures along the escarpment north of Lukachukai, Apache County, Arizona. The Lukachukai is the only member of the Wingate sandstone that is present at the type locality near Fort Wingate.

The lower Rock Point Member is generally more like the underlying Chinle Formation, being a slope-forming unit, rather than the cliff former widely recognized as the typical Wingate Sandstone in Utah. The unit contains fossil teeth and bone fragments of reptiles, believed to be of Late Triassic age, similar to modern-day crocodiles. The fine-grained character of the unit, and the thin, small-scale cross-bedding suggest that the unit was deposited by streams.

Although the Rock Point and the overlying Lukachukai members are closely related stratigraphically, practical considerations of general appearance and mode of origin would dictate placing the base of the Wingate at the base of the Lukachukai Member as in the type section. Anyway, Pipiringos and O'Sullivan realized that the J-0 unconformity separates the Rock Point Member from the overlying Lukachukai Member, thus providing a natural distinction with which to separate the two members. Consequently, the Rock Point is now considered to be a member of the Chinle Formation (Lucas and Hayden 1989; Dubiel 1989). The term "Lukachukai Member" then becomes synonymous with the name Wingate Sandstone, and is unnecessary for nomenclatural distinction of the impressive sandstone (Dubiel 1989).

The Wingate Sandstone (formerly Lukachukai Member) is a highly cross-bedded cliff-forming sandstone, no doubt of windblown origin. It

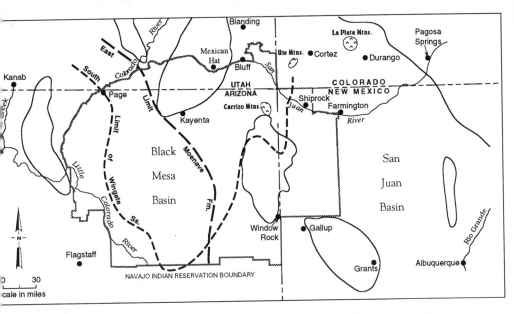

Figure 11 Represented by the Short-Dashed Line, the Approximate Southern Limits of the Wingate Sandstone, Formerly Known as the Lukachukai Member. Where present, the Wingate forms massive vertical cliffs that are formidable barriers to travel across the uplifts. The long-dashed line marks the eastern recognized limits of the Moenave Formation, a sandstone-siltstone unit that is partly equivalent to the upper Wingate Sandstone and the Kayenta Formation. These units, once believed to be of Triassic age are now known to be Jurassic-age deposits. Outlines of major geologic uplifts (see figure 2) are shown in light lines.

is a prominent, red, cliff former across most of Navajo Country. It extends east from the Echo Cliffs across northeastern Arizona and northwestern New Mexico as far as the Lukachukai Mountains and Beclabito dome in the Four Corners area, and north across eastern Utah (see figure 11).

The original type section of Dutton (1885) in the cliffs just north of Fort Wingate was later found to consist mostly of what is commonly mapped as the Entrada Sandstone in other parts of the Colorado Plateau. After heated debate among field geologists, Harshbarger and his co-authors (1957) restricted the term "Wingate" to the lower half of the originally described sandstone, retaining the site as the historical type section of the Wingate Sandstone. The J-0 unconformity at the base of the Wingate Sandstone (Lukachukai) is now believed to mark the base of the Jurassic System on the Colorado Plateau, making the Wingate Sandstone Early Jurassic in age (see figure 12). The more recent work of Maury Green of the USGS published in 1974, has shown that there is no Wingate Sandstone (Lukachukai) in the type section near Fort Wingate; the lower section referred to as Lukachukai by Harshbarger and associates is now called the Iyanbito Member of the lower part of the Entrada Sandstone. Ironically, that leaves the widely recognized and mapped Wingate Sandstone without a type section, theoretically an impossible situation by modern stratigraphic doctrine.

MOENAVE FORMATION

A distinctive unit only in western Navajo Country, west of the Defiance Uplift, the Moenave Formation is a reddish sandstone that overlies, and apparently interfingers with, the Wingate Sandstone. It is distinctive in that the formation consists of lenticular beds of sandstone, alternating between stream deposits and windblown dune sands in origin, that form blocky cliffs above the more massive and rounded cliffs of the Wingate. It forms the basal sandstone of the Glen Canyon Group along the Echo Cliffs monocline, west of the pinchout of the Wingate Sandstone, and interfingers eastward with the still younger Kayenta Formation (see figure 11). The Moenave Formation was named for exposures near Moenave, 8 miles west of Tuba City, Arizona, and is subdivided in that area into the lower Dinosaur Canyon Sandstone Member and the upper Springdale Sandstone Member, based mainly on different topographic expressions in the sandstone cliffs. A few fossil fragments of primitive crocodile-like reptiles, originally believed to be of Triassic age, have been found in the lower member.

KAYENTA FORMATION

Sandstones of the Kayenta Formation, deposited by ancient streams, dominate the stratigraphic position above the Wingate Sandstone over much of northern Arizona and southeastern Utah. The formation is important to the scenery of this region because it is a firmly cemented rock that supports the high, reddish cliffs of Wingate Sandstone throughout Canyonlands. The Kayenta, named for exposures just north of Kayenta, Arizona, overlies the Moenave Sandstones along the Echo Cliffs and grades laterally into Moenave beds toward the east. The Kayenta Formation is recognizable as a ledge-forming, relatively thin-bedded sandstone with medium- to small-scale cross-beds typical of stream deposits. Dinosaur tracks are common features on upper bedding surfaces. Although a prominent formation north and west of Black Mesa, the Kayenta is not present east of the eastern erosional margin of the Black Mesa structural basin, nor south of the immediate Four Corners region. It is not known to be present on the Defiance Uplift, nor

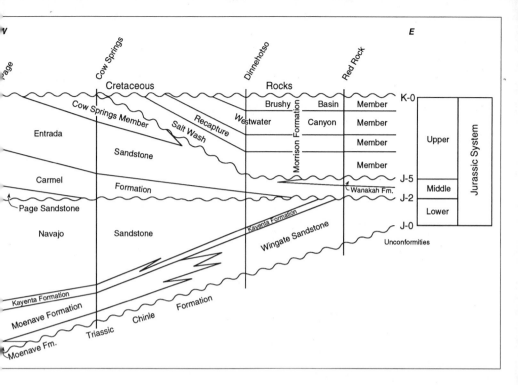

FIGURE 12 Diagrammatic Cross Section Showing the Relationships of the Various Named Units of Jurassic age with Intervening Regional Unconformities, Shown by Wavy Lines. The line of section is from west to east across northern Arizona.

eastward into the San Juan Basin of New Mexico. Kayenta beds grade to siltstones south of The Gap in the Echo Cliffs and south of Tuba City, making less distinctive exposures in southwestern Navajo Country.

NAVAJO SANDSTONE

Although the massive sandstone bodies of Jurassic age seem generally to come and go, grading into one another across Navajo Country, one formation, the Navajo Sandstone, remains distinctive. Named by Herbert Gregory in his landmark USGS Professional Paper in 1917 for exposures "in Navajo country," the Navajo Sandstone forms steep-walled rocky canyons capped with unique rounded, almost mammary, knolls of light-colored, highly cross-stratified sandstone that extend for hundreds of square miles across the slickrock terrain of western Navajo Country (see figure 13). Although usually a white or light gray color, the Navajo Sandstone locally becomes variably pink or light red for no apparent reason.

ENTER SAND

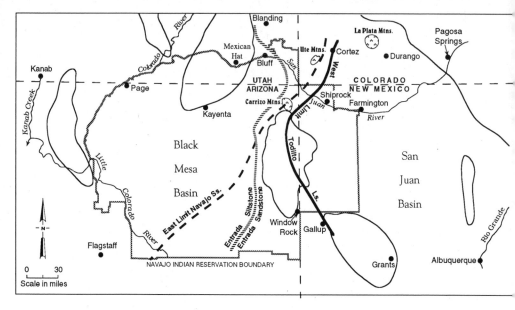

F I G U R E 1 3 The Eastern Limits (Heavy Dashed Line) of Recognizable Navajo Sandstone. The solid dark line represents the western limits of the Todilto Limestone (or Formation). The patterned line is a zone of facies change in the Entrada Sandstone; to the west the formation is a siltstone that is seen as softly weathering comical goblin-like spires and pinnacles; to the east, the formation forms massive cliffs of sandstone with numerous natural arches in southeastern Utah. All are of Jurassic age. Outlines of major geologic uplifts (see figure 2) are shown in light lines.

The formation consists of the cemented and well-preserved remains of windblown sand dunes that accumulated on a vast desert that would rival today's Sahara Desert. The quartz-sandstone deposits extend westward into present-day Nevada, where they are called the Aztec Sandstone, and northward into the Wasatch Mountains of central Utah, where they are called the Nugget Sandstone. Individual trough-shaped cross-bedding sets commonly range from 20–50 feet in thickness, and have been reported to reach as thick as 100 feet in northwestern Navajo Country. Thin beds of dolomite and chert, perhaps a foot or two in thickness, punctuate the monotony of the Navajo Sandstone locally in northern Navajo Country, believed to be the deposits formed in short-lived playa lakes scattered across the Navajo desert. The overall formation thins eastward from about 2,000 feet in Zion Canyon to about 1,000 feet near Navajo Mountain, continuing to thin to a featheredge west of the Defiance Uplift. The Navajo Sandstone is not present in exposures of Jurassic rocks in the Lukachukai and Chuska Mountains, northeast of the Four Corners area.

CHAPTER EIGHT

The rugged, inhospitable, nearly impenetrable country west of the southern Monument Upwarp, in the general region between Hoskinnini-Nakai-Skeleton mesas and Navajo Mountain to Glen Canyon, is made quite unfriendly by intricately eroded exposures of the Navajo Sandstone. This was a natural, huge hide-out for many Navajos who eluded Kit Carson's infamous roundup of the winter of 1863–64. Here, new or entirely isolated Navajo clans developed nearly separate subcultures while neighboring Navajos were forced on The Long Walk to the Fort Sumner Indian Reservation for four years of starvation and hardship. Even now, those living in this isolated corner of Navajo Country are far less Americanized and maintain more primitive living habits than their cousins to the east. This remains perhaps the most remote and little-known region in the conterminous United States today.

TRIASSIC-JURASSIC BOUNDARY

One of the most vexing problems in unravelling stratigraphic relationships in Navajo Country has been finding the boundary between the Triassic and Jurassic periods. The problem results from the rocks themselves, which are entirely of continental origin in this region and thus contain none of the usual fossils that are diagnostic of either geologic period. The rocks consist of endless square miles of sandstone that reach hundreds of feet in thickness, deposited on great, seemingly lifeless deserts, found in insurmountable cliffs; only a few trackways and burrow tubes mar the ubiquitous cross-beds. If a thin bed of shale or limestone can be found anywhere, it is void of traces of living things that can be used to distinguish time.

Seemingly more through intuition than on tangible evidence, Gregory assigned all of the rocks lying between the Chinle Formation and the Dakota Sandstone to the Jurassic System. If nothing else, that conclusion was handy for mapping purposes, and proved to be apparently correct.

Then, in 1931 a geologist, Barnum Brown, found fragments of fossil bone in rocks well above the Chinle that came from reptilian tetrapods (four-legged animals that lived something like crocodiles) in the Moenave Formation. He named these specimens of the previously undescribed primitive crocodile *Protosuchus richardsoni*. The material was compared with other vertebrate finds in South Africa and western Europe, resulting in an apparent Triassic age, but none of these rocks had been dated with any certainty. At first these fragmentary fossils were believed to be of Late Triassic age, and rocks below the Navajo Sandstone were assigned to the Upper Triassic. Since that time, small primitive dinosaur-like vertebrate fossils have been found in other parts of the Glen Canyon Group. Milton Wetherill of the pioneer Wetherill family who discovered Mesa Verde and other important archeological sites, found a small dinosaur in 1933 midway up in the Navajo Sandstone in the Kaibito Plateau area, which was assigned to the genus *Segisaurus*. E.F. Brady found another dinosaur in 1935 in the same area, which was similar to the genus *Ammosaurus*, then dinosaur tracks were found in the Kayenta Formation. Two skeletons of carnosaurs were found by S.P. Welles from the base of the Kayenta Formation near Moenave, causing him to assign the Kayenta to the Jurassic System. Meanwhile fossil fish of the genus *Semionotis* were found near the primitive crocodile find in the Moenave. Thus, fossil amphibians, dinosaurs, fish, freshwater clams, and snails have been recovered from various units, from the Moenave to the Navajo formations, all having Late Triassic to Early Jurassic affinities elsewhere in the world.

Of apparently greater significance in resolving the issue was the discovery of palynomorphs, fossil plant spores and pollen, throughout the Glen Canyon Group reported by Peterson and Pipiringos (1979). These plant fossils have been dated and verified as Early Jurassic in age by various palynologists, again placing the entire Glen Canyon Group down to the base of the Wingate Sandstone (Lukachukai) in the Jurassic System. A regional erosional surface, the J-0 disconformity, is now believed to separate the Rock Point Member (Triassic) from the (Lukachukai Member) Wingate Sandstone (Early Jurassic). Gregory was indeed correct, at least as far as can be determined at this time.

SAN RAFAEL GROUP

A section of largely red sedimentary rocks that overlies the Navajo Sandstone at the J-2 unconformity was named the San Rafael Group for exposures in the San Rafael Swell to the northwest of Navajo Country between Hanksville and Price, Utah. Although largely unfossiliferous, this section is considered to be of Middle Jurassic age. The originally defined group includes the Carmel and Entrada formations, that occur between the J-2 and J-3 unconformities, and the Curtis and Summerville formations that are between the J-3 and J-5 unconformities. Thus, the San Rafael Group lies between the J-2 and J-5 unconformities (see figure 12).

CARMEL FORMATION

A regional erosional surface, the J-2 disconformity, is found separating the Navajo Sandstone from the overlying Carmel Formation in western

Navajo Country. Like the Navajo, the Carmel Formation thins regionally from a western, mainly marine section in the vicinity of Zion Canyon to a thin red-bed section in western Navajo Country, only to thin to a pinchout west of the Defiance Uplift. In a small area around Page, Arizona in northwestern Navajo Country, a windblown sandstone, named the Page Sandstone occurs above the J-2 unconformity and below the Carmel. East of Page, as far as about the western flank of the Defiance Uplift at Chinle, only the upper member of the Carmel Formation occurs, consisting of dark reddish brown siltstone and mudstone that is relatively soft, only exposed in roadcuts and washes. At many places, such as Red Mesa and Bluff, Utah, the Carmel is highly contorted, having been crumpled by slumping and compaction as it was being buried by the overlying Entrada Sandstone. Thus, the Carmel Formation occurs in Navajo Country west of the Defiance Uplift as thin, soft red beds separating the Navajo Sandstone below from the Entrada Sandstone above.

ENTRADA SANDSTONE

As it is usually remembered, as in Arches National Park and the Moab region to the north, the Entrada Sandstone above the reddish brown Carmel Formation is a massive cross-bedded red sandstone that forms magnificent nearly vertical cliffs, often hosting natural arches. This massive sandstone facies, usually interpreted to be of windblown (eolian) origin, extends south through Canyonlands and southwestern Colorado onto the Defiance Uplift, cropping out along the western foothills of the Chuska Mountains and eastward along the northern flank of the Zuni Mountains (see figure 13). It forms the upper impressive cliffs north of Fort Wingate.

West of the crests of the Defiance and Monument uplifts the Entrada changes to a pinkish red siltstone and sandy mudstone facies that erodes to softly shaped slopes with intricate ghoulish erosional characters, such as at Baby Rocks east of Kayenta and east of Chilchinbito, and at the Valley of the Goblins north of Hanksville. This siltstone facies appears to have been deposited in relatively quiet waters west of the eastern windblown coastal dune fields in Middle Jurassic time (see figure 13). It can be difficult to distinguish from the underlying Carmel Formation, except for the lighter reddish hues and somewhat more massive bedding in the Entrada.

TODILTO LIMESTONE

Above the eastern massive facies of the Entrada Sandstone, a relatively thin limestone and gypsum formation is found throughout much of the

San Juan Basin. Named the Todilto Limestone (or Formation) for exposures in Todilto Park on the Defiance Uplift, the unit is a marker that clearly defines the top of the Entrada Sandstone. In the central San Juan Basin, as seen between Cuba and San Ysidro, New Mexico, the Todilto consists of gypsum that forms prominent white cliffs. Along the margins of the Todilto depositional basin, especially in the San Juan Mountains where it is called the Pony Express Limestone, and on the Defiance Uplift, as at the type section in Todilto Park, the formation is a thin limestone. When present in the section this unit, probably an inland lake deposit, makes it easy to separate the Entrada from overlying beds. However, recognizable Todilto beds are not present to the west and northwest of the Chuska Mountains, making some correlations at this stratigraphic level in western Navajo Country rather tentative (see figure 13).

COW SPRINGS SANDSTONE

A light-colored windblown sandstone, named the Cow Springs by Harshbarger, Repenning and Irwin (1957), is something of a maverick formation. By definition, the Cow Springs Sandstone overlies the Entrada Sandstone and may be an eolian equivalent of the Todilto Limestone, yet all of these units never seem to crop out together in any one section, and relationships are clouded. The type section of the formation is near Cow Springs Trading Post, southwest of Kayenta, where it is a nearly white, massive, highly cross-bedded sandstone (see figure 12). The prominent cliff-forming sandstone is best seen below the northern escarpment of Black Mesa and on the southern Defiance Uplift where it thickens dramatically southward from Todilto Park to Lupton. It was originally believed to be a windblown equivalent of parts of both the Summerville and lower Morrison formations, perhaps also the Todilto Limestone, and the approximate equivalent of the Bluff Sandstone, a basal windblown sandstone now considered to be a facies of the basal Morrison. If all of this sounds confusing, it is. That is because there are so many eolian sandstones of Jurassic age in Navajo Country. They all look much the same, and their lateral interrelationships are obscured by the nature of the scattered areas of outcrop. After years of study of Jurassic stratigraphy on the Colorado Plateau, Fred Peterson of the USGS now considers the Cow Springs to be a white facies of the Entrada Sandstone—a sensible interpretation that simplifies the entire matter.

Summerville Formation

Brown thinly bedded siltstones above the Entrada in the Four Corners area were originally mapped as Summerville Formation of the San Rafael Swell region. The unit formerly called Summerville north of the Four Corners area between Monticello and Moab, Utah, is now, however, considered to be the basal unit of the Morrison Formation, the Tidwell Member, because the true Summerville occurs below the J-5 unconformity, while the Tidwell Member is above that regional surface (see figure 12). The Summerville is correlated by Peterson of the USGS with the Moab Tongue of the Entrada Sandstone in Canyonlands as approximately age-equivalent, lying directly beneath the J-5 unconformity. What then are the red beds above the Entrada Sandstone in the Four Corners area? Peterson correlates these reddish brown siltstones and sandstones with the Wanakah Formation of southwestern Colorado, largely age-equivalent to the Curtis Formation, below the typical Summerville of the San Rafael Swell. Differences in rock type and the differing ages of the two red-bed formations in question, warrant calling the brown silty beds above the Entrada the Wanakah Formation in the Four Corners area; the formation is especially noticeable between Red Mesa and Bluff, Utah. Thus, brown siltstones lying between the Entrada and the Morrison formations, formerly mapped as Summerville, are now known as the Wanakah (pronounced "Wah-nah!-ka") Formation of Middle Jurassic age (Condon 1989).

UPPER JURASSIC

Morrison Formation

Generally above the massive Lower and Middle Jurassic windblown sandstones and above the J-5 unconformity, drab-colored sandstones and shales of the Morrison Formation dominate the landscape, especially in the Four Corners area of Canyonlands Country. Although the name was derived from exposures east of the Rocky Mountains near Morrison, Colorado, the formation is widespread across the Colorado Plateau. The Morrison Formation consists of four members in Navajo Country, not all of which are present at any one locality. These are, in ascending order, the Salt Wash, Recapture, Westwater Canyon and Brushy Basin members. In recent years, the Bluff Sandstone and the Tidwell Member have been considered as basal members of the Morrison as well. All members contain both sandstone and mudstone, but alternate in relative amounts of each, forming a generalized stair-step topography in areas of exposure. All members are interpreted to have been deposited by streams and in associated lakes in Late Jurassic time, except for the Bluff Sandstone that is an eolian facies of the Salt Wash Member. These deposits thin by erosional truncation at the base of the overlying Cretaceous Dakota Sandstone southward across Navajo Country, with the Dakota overlying older formations to the south. The Morrison has been studied extensively on the Colorado Plateau because its sandstones in many places contain important deposits of uranium. Although no quarries have been developed in Navajo Country, the Morrison Formation is well known to contain fossil dinosaurs which are quarried at several locations to the north. One small quarry, opened near San Ysidro, New Mexico, in *Dinétah*—Old Navajoland—produced about 20 percent of a *Camarasaurus* skeleton, a large plant-eating dinosaur, according to J. Keith Rigby, Jr. (1982). Teeth of a meat-eating dinosaur, *Allosaurus*, were also found in the quarry. Since the fossil teeth were well worn, scientists speculated that the *Allosaurus* probably died while feeding on the *Camarasaurus* carcass.

Considered originally as the youngest formation in the San Rafael Group, the Bluff Sandstone is now believed to be a basal windblown sandstone facies of the Salt Wash Member of the Morrison Formation. The name was derived from Bluff, Utah, just north of the San Juan River, where cliffs of the massive, highly cross-bedded sandstone are prominent. It forms the upper cliffs in Red Mesa near the Arizona-Utah border in the Four Corners area and seems to thicken northward, filling an ancient topographic depression in the Blanding Basin of southeastern Utah. Originally believed to be an extension of the Cow Springs Sandstone, which occurs below the J-5 unconformity, it is now seen as a separate accumulation of dune sands in a distinctive isolated basin of deposition above the J-5 surface (see figure 12). As such, it is a lateral equivalent of the lower parts of the Salt Wash Member of the Morrison Formation. Deposits in west-central New Mexico, north of the Zuni Mountains, have been given local names in New Mexico north of the Zuni Uplift by Clay Smith because of uncertain relationships with the widely recognized named units in the Morrison to the north. However, this local terminology has not been widely accepted by geologists.

The lower, regionally recognized member of the Morrison Formation is the Salt Wash, a ledgy cliff-forming unit dominated by fluvial sandstones but containing interbedded mudstones. The member is exposed extensively in the greater Four Corners area, north of the Defiance Uplift, and thins to zero southward along the eastern flank of the Chuska Mountains. North of Navajo Country, the Salt Wash Member constitutes the lower half of the Morrison Formation in eastern Utah and western Colorado. Exposures in the Four Corners and to the north commonly contain a variety of uranium minerals, forming marginally economic ore deposits at times when that substance is marketable.

Fluvial sandstones of the Salt Wash Member interfinger upward and southward with a dominantly mudstone facies of the Morrison called the Recapture Member. This member thickens southward along the Chuska Mountains at the expense of the Salt Wash Member to become the lower unit of the Morrison along the north flank of the Zuni Mountains. Drab-colored mudstones of the Recapture contain thin fluvial sandstone stringers, especially in areas of interfingering with the Salt Wash. The Recapture Member interfingers southward from Navajo Country into sandstones, sometimes called the Thoreau south of the Defiance and Zuni uplifts.

Above the shaly Recapture Member, stream-deposited sandstones, often conglomeratic, again form ledgy cliffs and dominate the middle part of the Morrison Formation from the Four Corners–Blanding Basin area southward to the Zuni Mountains. The sandy middle Morrison is known as the Westwater Canyon Member. It is the Westwater Canyon and equivalent beds that were highly productive of uranium ores in the Grants-Ambrosia Lake mining district.

The upper, dominantly mudstone section of the Morrison Formation is generally known across the Colorado Plateau as the Brushy Basin Member. It forms soft, gently rounded slopes of drab-colored mudstone, usually gray with shades of dull green, pink and brown, that can produce treacherous road conditions when wet. The upper half of the typical Morrison Formation north of Navajo Country is the Brushy Basin Member. A contained fluvial sandstone complex, known locally as the Jackpile Sandstone, produced uranium ores in the North Laguna mining district.

SUMMARY

Continental environments of deposition prevailed across Navajo Country throughout Triassic and Jurassic times. Tidal flats of the Early Triassic Moenkopi Formation west of the Monument and Defiance uplifts soon were replaced by stream and lake deposits of the Chinle Formation, known in western Navajo Country as the Painted Desert. Then desert conditions dominated the landscape, as vast accumulations of dune sands buried Navajo Country under hundreds of feet of wind-blown deposits in Early to Middle Jurassic time, forming the Wingate, Navajo, and Entrada/Cow Springs sandstones, punctuated locally by lesser red-bed deposits. Finally, in Late Jurassic time, Navajo Country returned to being extensive continental lowlands dominated by meandering streams and associated lakes that distributed sands and mud to form the widespread Morrison Formation.

These often colorful and rugged terrains of exposed Triassic and Jurassic rocks are found in broad belts ringing uplifts. They are not present across the structurally higher regions, as erosion has removed all but the older, more resistant Paleozoic rocks on the Nacimiento, Zuni, Defiance, Kaibab and Monument uplifts. Triassic and Jurassic strata underlie the San Juan and Black Mesa basins, but there they are buried beneath the younger Cretaceous and Tertiary cover. Consequently, it is along the flanking structural shelves, between uplifts and basins, that these complex and fascinating sedimentary rocks dominate the imagination.

ROCKS OF THE BASINS
CRETACEOUS-TERTIARY TIME

N ot much was going on in Navajo Country as the Jurassic Period gave way to Cretaceous times. Deposits of the Morrison Formation formed the surface of the land for quite some time. Then, particularly to the north, in eastern Utah and western Colorado, streams again began to flow, this time mostly reworking the weathered surface to form what is called the Burro Canyon Formation, conglomeratic stream deposits that only barely reached the Four Corners area. If there was any geologic activity south of the present-day San Juan River in Navajo Country in Early Cretaceous time, the deposits did not survive or have not been recognized.

To the west and northwest, however, times were changing. The seaway that had prevailed in what we now call Nevada and western Utah began to dry up as great compressional forces in the Earth's crust began crumpling the western margin of the continent in Late Jurassic and Early Cretaceous time. Large-scale folding and thrust-faulting began to shorten the continental margin, pushing up mountain ranges where seas once had prevailed, forcing the seas to seek other basins. It all began in the west, as granitic intrusive igneous rocks formed in what is now the Sierra Nevada of California, only to migrate slowly eastward as Jurassic time progressed to arrive in central Utah by the Early Cretaceous. This great west-to-east compressional event marks what plate-tectonics theorists consider to be the initial but progressive subduction of the Pacific oceanic plate beneath the western margin of the North American continental plate. Erosion attacked the newly formed lands, and sediments— mostly boulders and sand—were shed eastward, eventually accumulating to thousands of feet of conglomerate along what is now the western margin of the Colorado Plateau.

By Late Cretaceous time, the main seaway had shifted eastward into the Midcontinent as mountain ranges sealed off the western shorelines. Finer-grained sediments, the sand and mud, washed eastward into the expanding Cretaceous seaway to become trapped along shorelines as beach sands. The finer material drifted into deeper waters to settle as marine mud. As waters deepened, and the Cretaceous seaway expanded, shorelines migrated westward, encroaching on lands that would become the Colorado Plateau, utilizing sediments derived from the now-mountainous west.

ROCKS OF THE SAN JUAN BASIN

Because later erosion has effectively tried to level the landscape, rocks of Cretaceous age are preserved only in structurally depressed areas of Navajo Country. Although they must have originally covered much if not all of the land, rocks of this younger age are now preserved only in the San Juan Basin to the east and the Black Mesa Basin to the west. In the best Navajo tradition, this discussion will begin in the east, where the San Juan Basin contains the thickest and best-known rocks of Cretaceous age.

DAKOTA FORMATION

The initial shoreline of this continental interior sea reached the Four Corners area in the early part of Late Cretaceous time. At first the results were reworking of the weathered sediment surface into shoreline deposits, but as time progressed more sediments were incorporated from the western sources. Advancement of the shoreline was slow but not particularly steady, and as fluctuations in sea level occurred, beaches waxed to and fro across the land. Where beach sands would first accumulate, coastal swamps developed when a retreat of the shoreline exposed coastal areas to swampy lowlands, resulting in much carbonaceous material being incorporated into the coastal deposits and coals formed locally. Finally beach sands overwhelmed the former lands as the sea continued its relentless inundation of Navajo Country.

The deposits of this advancing Cretaceous shoreline are called the Dakota Formation, named for similar sandstones that occur in the northern Midcontinent. The Dakota Formation, consisting of nearshore marine sandstones and interbedded carbonaceous mudstones and coal beds, was deposited over the Burro Canyon sedimentary rocks to the north, generally on rocks of the Morrison Formation in Navajo Country, but progressively overlying older rocks to the south. As the formation is well-cemented sandstone for the most part, it is resistant to erosion. It underlies vast upland regions, such as the Great Sage Plains north of the San Juan River, caps plateaus and mesas, such as Toh Atin

Mesa along the Utah-Arizona border west of the Carrizo Mountains, and forms great hogback ridges that ring the margins of the San Juan Basin, such as those seen east of Window Rock and north of Gallup.

MANCOS SHALE

Coastal sandy deposits formed the beaches of the Cretaceous sea, as the finer-grained sediments, silt and mud, were washed into deeper, quieter waters offshore. Marine waters deepened as the sea floor slowly sagged, forming environmental conditions that were conducive to the accumulation of hundreds to thousands of feet of marine mud. These deposits were named the Mancos Shale for exposures of the nearly black, softly rounded slopes in the Mancos River Valley of southwestern Colorado. As the shoreline of the Mancos sea reached its maximum extent, some 200 miles south and southwest of Navajo Country, a limy deposit known as the Greenhorn Limestone Member became widespread in the lower Mancos Shale across the Four Corners region. That part of the Mancos Shale lying below the Greenhorn Limestone is sometimes called the Graneros Shale Member. Black shales of the Mancos average more than 2,000 feet in thickness, and in places such as near Grand Junction, Colorado to the north, they reach thicknesses greater than 3,000 feet. Dark-gray to black colors in the Mancos Shale result from reducing (not oxidizing) environments of deposition, where organic matter is readily preserved and iron-rich sediments easily revert to ferrous, low-oxygen minerals. Such organic-rich sedimentary rocks make ideal source beds for petroleum, and when crushed and heated, Mancos Shale will indeed release small amounts of oil.

Being relatively nonresistant to erosion, exposures of Mancos Shale weather to broad open valleys and smooth gentle slopes except where capped by more resistant sandstones of the overlying Mesaverde Group. Because of these valley-forming characteristics, Mancos Shale outcrops provide natural surfaces for town and highway construction, especially in the Four Corners area where broad expanses of the Four Corners Platform are underlain by the black shale. The same quality that provides broad, nearly barren valleys—the high clay content of the Mancos—also turns the land to gumbo when wet, making highway, airport and town sites bowls of gooey muck that ensnare roads and buildings with little warning.

Although the mud-size sediments were deposited into accumulations of oxygen-poor, black, smelly, bottom mud, waters of the Mancos sea were apparently well oxygenated, as marine sea life flourished. Fossil mollusks—clams, oysters, and cephalopods—are commonly

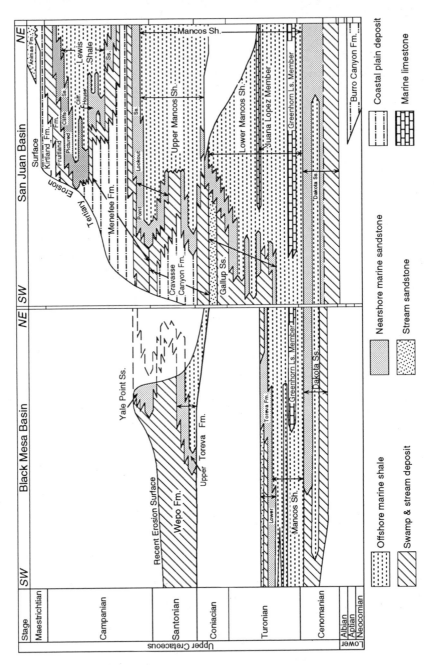

FIGURE 14   Correlation Chart Showing Formation Names and Age Relationships Between Rocks of Cretaceous Age in the Black Mesa and San Juan Basins in Navajo Country. Stage names along the left side are internationally accepted names for time intervals within the Cretaceous period. Blank areas are time intervals for which no rocks are present; in other words, they represent unconformities. Simplified from Peterson and Kirk (1977)

CHAPTER NINE

found in exposures of the Mancos Shale, as are shark teeth. Of these, the ammonitic cephalopods have been used extensively for dating and correlating the rocks over broad regions of the continent. The tiny single-celled animals called foraminifers have produced millions of microscopic fossil shells found abundantly in the Mancos Shale that are even better tools for correlation, as they are sufficiently small to be found by the thousands in well cuttings and cores.

Mesaverde Group

After what must have been several million years, the open Mancos sea began to shrink, and shorelines backed eastward from the region toward the main continental seaway. This was not a simple process, as shoreline sandstone deposits marched back and forth repeatedly across the San Juan Basin as Cretaceous time progressed, forming complex stratigraphic relationships that after years of study still give geologists nightmares. Myriad formation and member names resulted, as the various deposits come and go both laterally and vertically with time (see figure 14).

First there was the Gallup Sandstone, with several stratigraphic complexities that give rise to locally derived names of members and tongues as conditions dictate. The Gallup is best developed along the northern flank of the Zuni Mountains and into the subsurface of the southern San Juan Basin. Laterally developed coal beds, such as those being strip-mined near Window Rock, formed periodically in landward swamps adjacent to the sand beaches, as the pulsating shoreline episodically retreated northeastward. A time of regional non-deposition and/or erosion punctuates the section after deposition of this first hint of regional regression of the sea.

Best-known exposures of regressive (caused by lower sea level) and transgressive (formed as sea level was rising) shoreline deposits of the Late Cretaceous seas are found within and surrounding Mesa Verde in southwestern Colorado. There the lower regressive sandstone, named the Point Lookout Sandstone for a promontory overlooking the entrance to Mesa Verde National Park, forms prominent cliffs above the dark-gray steep slopes of the Mancos Shale. Swampy conditions developed landward (southwest) of the Point Lookout beach environment where carbonaceous shales and coal accumulated in the Menefee Formation (see figure 14). Coal from the Menefee has long been mined for local use in the Durango area—in older times for household heating purposes and for fueling the smelter, now mainly for fueling the steam locomotives used on the Durango and Silverton Narrow Gauge Railroad.

Once again the shoreline migrated southwestward, and shoreline deposits of the Cliff House Sandstone were deposited over the Menefee. The Cliff House Sandstone received its name for the many Anasazi cliff dwellings that are nestled beneath the massive sandstone cliffs in Mesa Verde National Park. These three units, the Point Lookout Sandstone, the Menefee Formation and the Cliff House Sandstone (in ascending order) form the classic Mesaverde Group. As the shoreline retreated for the final marine episode, Late Cretaceous beaches retreated back to the great inland seaway, the event marked by the Pictured Cliffs Sandstone. Offshore black shales, known as the Lewis Shale, almost identical to the Mancos Shale, were being deposited in open waters to the northeast. Through all of these fluctuations of conditions, offshore bars were constructed locally adjacent to shoreline deposits, each receiving local stratigraphic names.

All of this is not only complex, it is confusing to everyone who has not spent years studying the Cretaceous rocks of the San Juan Basin. There is now far more written about the idiosyncracies of these deposits than most of us care to know about. The accompanying diagram (see figure 14), borrowed from Peterson and Kirk (1977), demonstrates the complexities and the associated rock names applied to the Cretaceous System of the San Juan Basin, perhaps better than words can describe. For a more detailed summary of these stratigraphic relationships, the summary papers (Molenaar 1977 and 1983) are recommended reading.

POST-MESAVERDE ROCKS

With the final withdrawal of the sea from Navajo Country, sedimentary deposits of continental origin again dominate the section. Lagging closely behind the retreating Pictured Cliffs beach deposits were thick coals that form the base of the Fruitland Formation. These coal beds, the thickest in the San Juan Basin, are strip-mined to fuel the Four Corners Power Plant east of Shiprock. Exposures of the coal may be seen in roadcuts along U.S. Highway 550 south of Durango, Colorado and in the Bisti badlands south of Bisti Trading Post. Above the Fruitland coal, stream deposits of gray mudstones and sandstones constitute the Fruitland-Kirtland Formations deposited on the emerging lowlands left stranded by the retreating Cretaceous sea. Fragmentary dinosaur fossils from the Fruitland-Kirtland Formations indicate a latest Cretaceous age for the deposits (Hunt and Lucas 1992). Finally, as volcanism and uplift began in southwestern Colorado in latest Cretaceous time, purplish conglomerates—sandstone and mudstone of the McDermott Formation—were shed by streams onto plains bordering the northern San Juan Basin. These form prominent colorful cliffs south of Durango. These local deposits of the McDermott Formation are considered to be the uppermost sedimentary rocks of Cretaceous age in the basin.

Laramide Orogeny

The land of the San Juan Basin was "high and dry" following the demise of the last dinosaurs of the Fruitland-Kirtland Formations and the consequent beginning of Tertiary time. Uplift of the Rocky Mountain West had begun with the encroachment of massive compressional forces of the Laramide orogeny from the west. The main results of this wave of mountain-building forces in the Earth's crust were a general uplift of the land, the accentuation of previously formed structures to produce the great monoclines of the Colorado Plateau, and, more specifically, the uplift of the San Juan Dome to the north of the San Juan Basin. With the gradual rise of this mountainous massif in southwestern Colorado, erosion began levelling the landscape, as it is often wont to do, and stream deposits flooded the structural San Juan Basin to the immediate south.

Tertiary Rocks

The first of the Tertiary deposits, the Animas Formation, consists of conglomeratic coarse sandstones, greenish gray in color, that were derived from intrusive igneous rocks, perhaps in the vicinity of the present-day La Plata Mountains. The coarser clastic sediments were dumped near the foot of the uplift, along the northern flank of the San Juan Basin, seen just south of Durango, and finer-grained sediments drifted on southward, eventually to interfinger with the Ojo Alamo Sandstone and the overlying Nacimiento Formation of the southeastern San Juan Basin. This sequence of deposits is of Paleocene (earliest Tertiary) age, as demonstrated by the fossil mammals it contains (Williamson and Lucas 1992). The Animas Formation forms the drab but varicolored, soft-weathering shaly hills along New Mexico Highway 44 from Aztec south to near Cuba, New Mexico.

As uplift of the structural San Juan Dome hastened and topographic relief became more pronounced, coarse sandstones of the San Jose Formation filled the San Juan Basin with several hundred feet of fluvial deposits. These include a rich, although fragmentary, fossil fauna of primitive plant-eating mammals *(Meniscotherium)*, hippo-like creatures *(Coryphodon)*, the earliest horses *(Eohippus)*, turtle shell fragments and crocodile teeth of Eocene age (Lucas and Williamson 1992). Deposits of the San Jose Formation form the cliff and bench scenery along the lower Animas River Valley and eastward from Aztec to near Dulce, the Gobernador area, in the deepest structural depression of the basin. The wooded ledgy and cliffy mountainous terrain, incised by deep gorges, such as Largo and Gobernador canyons, forms the heart of *Dinétah*, Old NavajoLand. There was the domain of the legendary Holy People, the home of Changing Woman and White Shell Woman, the birthplace of Monster Slayer and Born of Water, the monster-killing twins of Navajo myth. Although this region is of primary importance to Navajo mythology, it is not included in the Navajo Indian Reservation.

BLACK MESA BASIN

The western major structural basin, west of the Chuska Mountains and the Defiance Uplift, dominated by Black Mesa, is known to geologists as the Black Mesa Basin. Like its bigger cousin, the San Juan Basin to the east, the Black Mesa Basin contains the prominent remains of Cretaceous-age rocks, preserved from recent erosion by their lower structural position. Unlike the San Juan Basin, however, rocks of Tertiary age have been stripped from the high plateau country, leaving Cretaceous sandstones as caprock. Here, Hopis built their villages atop the high mesas, probably migrating from abandoned Anasazi dwellings of the Four Corners area—here near the heart of *Diné Bikéyah*, Navajo Country.

For generations, Hopis and Navajos have coexisted on Black Mesa; *Diné* graze herds within sight of the pueblo villages, Hopis watch from above. This coexistence has not been a happy one for either culture. Navajos believe that their rights to ancestral grazing lands is being jeopardized by Hopi activities; Hopis claim full rights to the land by first occupancy. Not only surface-use rights have been in open conflict; mineral rights have also been disputed since the time such subsurface treasures were realized. An Executive Order Reservation for the *Moqui* (Hopi) Indians was established in 1882, but a 1962 judicial decision divided the disputed lands into an Exclusive Hopi Reservation lying within the Navajo-Hopi Joint Use Area (Goodman 1982, 57). Fences have recently been built according to legal rulings by Federal judges, and Navajos have been, sometimes forcefully, relocated. New additions to the Hopi Indian Reservation north of Flagstaff, Arizona and west of the Navajo Indian Reservation boundary were announced in late 1992; settlement of the land dispute and associated additions to the Hopi land position have not been approved as yet. Exploratory drilling for possible petroleum wealth has been made impossible due to these land disputes since interest first materialized in the 1950s. Consequently, the petroleum potential of this region remains unknown pending future resolution of the conflict. Most of the minable coal reserves in Cretaceous rocks of Black Mesa lie north of the disputed territory, but some occur within the lands of legal "Joint Use."

The Cretaceous stratigraphic section seen in Black Mesa is similar in appearance to the Dakota-Mancos-Mesaverde section of the San Juan Basin. However, the ages of the principal players are somewhat different in the Black Mesa Basin from those of their eastern counterparts. Deposition of the first transgressive rocks of the Dakota Sandstone began somewhat later in the Black Mesa region as it took a while for the shoreline to migrate this far to the west. Deepening of the waters to allow mud deposition of the Mancos Shale also lagged a bit in time. Nevertheless the first signs of regression of the Late Cretaceous seas began somewhat earlier in the west, prompting the retention of different names for the rock units.

Although the generally regressive sandstones of the Toreva Formation of Black Mesa are similar in appearance to the Point Lookout Sandstone, the time of deposition was approximately coeval with that of the Gallup Sandstone of San Juan country. The disconformity, time of non-deposition, that terminates Gallup deposition to the east, lies within the Toreva in Black Mesa, making the upper sandstone member of the Toreva equivalent in age to the upper Mancos Shale of the San Juan Basin.

Coal-bearing shales of the Wepo Formation, similar in appearance and rock type to the Menefee Formation of Mesa Verde, are actually older and age-equivalent to the transition from Mancos Shale to the Point Lookout Sandstone at Mesa Verde. The uppermost sandstone in Black Mesa, the Yale Point Sandstone, was being deposited at about the same time as the Point Lookout Sandstone in the San Juan Basin. Although there is no doubt they were once deposited, no Cretaceous rocks above the Yale Point Sandstone have been preserved in Black Mesa country.

The lack of continuous exposures of Cretaceous rocks, and the extremely complex history of transgressive-regressive deposition, makes it impossible to relate precisely rocks of Black Mesa with those of Mesa Verde. The variations in time of deposition of the various rock units as described above is made possible by studying the evolutionary trends of the fossils, especially the foraminifers and the ammonitic cephalopods. The chart (see figure 14) showing the most likely relationships between Cretaceous rocks of Black Mesa and those of the San Juan Basin is simplified from that published by Peterson and Kirk (1977, 168–69).

TERTIARY ROCKS

Although none are present on Black Mesa, rocks of Tertiary age occur at the surface in the Little Colorado River drainage basin to the southeast in the vicinity of the Hopi Buttes. The section, named the Bidahochi Formation, has been divided into a lower member, consisting largely of lake sediments, a middle volcanic unit, and an upper sedimentary sequence of mainly stream-deposited sandstones and some shales. The lower member contains mudstones and muddy sandstones that form banded gray, brown and pink slopes and ledges. This is overlain by lava flows and related detritus, especially in the immediate vicinity of the Hopi Buttes volcanic field. The upper member is mainly composed of sandstones, some of which are lake deposits but most are of fluvial origin. They are poorly cemented to form rounded slopes separated by soft ledges. The Bidahochi Formation is a poorly exposed, generally nondescript slope-forming unit lying south of Ganado along the southwestern flank of the Defiance Uplift and south of Sanders but west of the Zuni Mountains along the Arizona–New Mexico border.

It has been postulated that the Bidahochi Formation may represent deposits of the ancestral Colorado and/or San Juan rivers drainage basin, prior to the time that the Colorado breached the Kaibab Uplift to empty into the Gulf of California. More reasonable interpretations indicate that the sediments were probably derived from surrounding local uplifted areas, such as the Defiance and Zuni uplifts, and deposited in and around a lake centered in the Hopi Buttes area.

Age of the Bidahochi Formation has been established as probably Pliocene (Late Tertiary) on the basis of vertebrate fossils, mainly small camels, mastodons, fox, beaver, rabbit, wolverine, large cat, reptiles, birds and fish, along with a few invertebrates, plants and pollen (Repenning, Lance and Irwin 1958, 128–29).

VOLCANIC NECKS

It was just here that Monster Slayer, born of Changing Woman, was plucked from the gray, barren plain by Bird Monster, and deposited on a ledge just beneath the peak of *Tsé bit' a'i*, the Rock With Wings

SHIPROCK   This classic diatreme (explosive volcanic vent) stands guard over the Four Corners region in northwesternmost New Mexico. The volcanic neck and associated radiating igneous dikes have intruded into the Mancos Shale of Cretaceous age as seen in this aerial view. The Chuska and Lukachukai Mountains rise on the horizon to the southwest.

(Shiprock). From the high ledge, Monster Slayer killed the adult male Bird Monster, who prayed on the men to feed his offspring, with an arrow of sheet lightening that his father, the Sun, had given him. Then, with another arrow of sheet lightening he killed the adult female Bird Monster, who was known to prey on women. Finally, he swirled the two infant Bird Monsters around his head four times each to create first an eagle and then an owl to benefit future generations of the five-fingered people. Monster Slayer was soon rescued from his high perch and lowered to the ground several hundred feet below by Bat Woman (Zolbrod 1984, 231–42).

Thus, numerous black, one might call ugly, spires found scattered across Navajo Country have affected Diné legend. Shiprock is but one. There are dozens of these distinctive volcanic necks here: Agathla south

AGATHLA PEAK Called El Capitan by Kit Carson, this peak rises above the desert south of Monument Valley near Kayenta. It is the solidified neck of a Tertiary-age explosive gaseous volcano (diatreme) that ripped its vent upward through the Triassic Chinle Formation. The Navajo name, *Agathla*, literally means "much wool."

of Monument Valley, Chaistla and Church Rock near Kayenta, The Beast north of Fort Defiance at the village of Navajo; Alhambra Rock south of Mexican Hat, the Mule Ear diatreme on the San Juan River, several prominences named locally as Black Rock, and a whole swarm of features known collectively as the Hopi Buttes in the southern Black Mesa Basin, to name a few.

Although generally known as volcanic necks, these are in reality the throats of highly explosive, gaseous, violent volcanic eruptions that rocked Navajo Country during Pliocene (Late Tertiary) time. Techni-

CHURCH ROCK   The eroded neck of an explosive volcano (diatreme), this geological formation rises east of Kayenta, Arizona. The Comb Ridge monocline, consisting of upturned beds of Navajo Sandstone (Jurassic), forms the southeastern margin of the Monument Upwarp to the left.

cally they are called diatremes. Most of the diatremes stand as variously shaped rock protuberances above the surrounding landscape because they are more highly resistant to erosion than is the surrounding country rock. The most prominent of these are Shiprock and Agathla, both rising hundreds of feet from their generally circular bases.

The diatremes consist mostly of breccias containing rocks brought up from the depths below, even from the Earth's mantle, with only minor amounts of intermixed dikes and sheets of basaltic, biotite-rich igneous rocks called minette. Some of the material within the diatremes consists of a microbreccia, sometimes called kimberlite. These often contain tiny garnets, best found in anthills, but none has yielded diamonds such as are found in the kimberlite pipes of South Africa and elsewhere. Most of the rock fragments, ranging in size from pebbles to

ROCKS OF THE BASINS   81

THE BEAST   This haunting formation, The Beast, the neck of an explosive volcano (a Tertiary-age diatreme), pierces the Cow Springs Sandstone (Jurassic) at the sawmill center of Navajo north of Fort Defiance.

several feet in diameter, consist of granite and metamorphic rock fragments brought up from the basement, mixed with sandstone, shale and limestone blocks broken from layers of Paleozoic rock through which the vent tore its way upward. However, in some, such as the Mule Ear diatreme, just south of the San Juan River where it crosses Comb Ridge, enormous blocks—up to the size of houses—of rocks that formed the surface at the time of eruption, later fell back into the eruptive chasm (see figure 15). There is such an array of strange rock types at Mule Ear diatreme that it was originally mistakenly mapped as glacial moraine. Many of the diatremes are intimately associated with dikes, some of which radiate from the central vent such as at Shiprock (Fitzsimmons 1973, 106–9).

The diatremes of Navajo Country often occur in linear belts, more or less superimposed on or along the great monoclinal flexures. Examples are found from Shiprock south along the Hogback monocline in northwestern New Mexico, and several occur between the Mule Ear

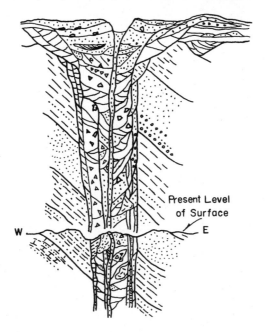

FIGURE 15 Diagrammatic Cross Section of a Typical Diatreme. These violently explosive volcanic vents are found in abundance in Navajo Country. Such vents are filled with rocks of all kinds carried up from great depths in the Earth's crust by eruptive gasses, mixed with rock fragments from shallower strata that have fallen back into the crater following eruptions. The higher structures in the diagram represent diatremes of the Hopi Buttes area where later erosion has modified the volcanic features only slightly; the surface deeper in the diagram represents a deeply eroded neck such as the one at Mule Ear near the San Juan River. From Ellingson (1973)

diatreme and Church Rock along the Comb Ridge monocline. These have apparently erupted along major basement fault zones that localized the monoclines and later the diatremes along areas of fundamental crustal weakness. Others, especially the Hopi Buttes of the southern Black Mesa Basin, defy such simple explanations, as they occur apparently randomly with little or no recognizable structural roots.

Certainly Navajo Country was a thunderous nightmare in Pliocene time, as vent after vent erupted with vengeance. The explosive and disastrous eruption of Krakatoa in historic times is probably indicative of each of the many blowouts that occurred here in the late Tertiary. The underlying cause of these explosive eruptions remains unknown—only the dark, ominous remnants mark their once-noisy existence.

LAND OF WHITE SPRUCE

It is little wonder that the upland forming the backbone of Navajo Country has also been the center of Navajo culture since the time *Diné* migrated westward from *Dinétah*, perhaps at the insistence of marauding Utes and Comanches. The climate is idyllic in summer in the high ranges, grasslands for grazing herds abound, and game is abundant, or at least has been. Canyons to the west provide shelter from the elements. Lush crops—especially of corn and, in the olden days, peaches—were protected from the severe winter season. In the Navajo language, the name for the highlands is *Chosgai*, now called *Chuska*, or "land of white spruce." It would seem as though this land inspired "The First Song of Dawn Boy":

> *Where my kindred dwell,*
>   *There I wander.*
> *The Red Rock house,*
>   *There I wander.*
> *Where dark kethawns [sacred sticks]*
>   *are at the doorway,*
>   *there I wander.*
> *At the yuni [seat of honor] the striped*
>   *cotton hangs with pollen.*
>   *There I wander.*
> *Going around with it.*
>   *There I wander.*
> *Taking another, I depart with it.*
>   *With it I wander.*
> *In the house of long life,*
>   *there I wander.*
> *In the house of happiness,*
>   *there I wander.*
> *Beauty before me,*
>   *with it I wander.*

Beauty behind me,
* with it I wander.*
Beauty below me,
* with it I wander.*
Beauty above me,
* with it I wander.*
Beauty all around me,
* with it I wander.*
In old age traveling,
* with it I wander.*
On the beautiful trail I am,
* With it I wander.*
—FROM N. SCOTT MOMADAY,
BETWEEN SACRED MOUNTAINS

This region is generally referred to as the Defiance Uplift, a prominently high combination of the Defiance Plateau and the Chuska Mountains that lie along its eastern margin. The highland is bordered on the east by the Defiance monocline, a sharp fold in the layered rocks that droops the geologic formations down into the Chuska Valley and San Juan Basin toward the east. The monocline lies generally north-south along the Arizona–New Mexico border, but in detail it is found to form a sinuous line. That is because the apparently northerly highland consists in detail of several structural elements—folds that trend northwest-southeast in keeping with the dominant structural fabric of the Colorado Plateau.

STRATIGRAPHIC RELATIONSHIPS

PERMIAN ROCKS

The Defiance Uplift seems to have been doomed to be a highland from the beginning. Modern-day erosion has revealed that red beds of Permian age lie directly on the Precambrian basement in Black Canyon south of Hunters Point and on Precambrian quartzite in the upper reaches of Bonito Creek, known as Quartzite Canyon, and in Blue Canyon northwest of Fort Defiance. Rocks of Cambrian through Pennsylvanian age are missing here, although they are present deep beneath the surface surrounding the uplift. Wells drilled in adjacent areas off the uplift show that rocks of Cambrian through Mississippian age lap onto the flanks of the structure from the north and west, and that the uplift was a source of sands deposited in the adjacent Pennsylvanian seaways.

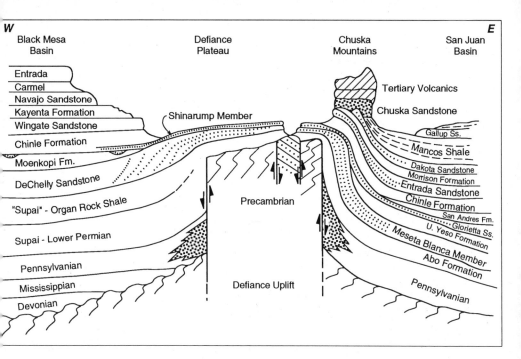

W
Black Mesa Basin | Defiance Plateau | Chuska Mountains | San Juan Basin

Entrada
Carmel
Navajo Sandstone
Kayenta Formation
Wingate Sandstone
Chinle Formation
Moenkopi Fm.
DeChelly Sandstone
"Supai" - Organ Rock Shale
Supai - Lower Permian
Pennsylvanian
Mississippian
Devonian

Shinarump Member

Precambrian

Defiance Uplift

E

Tertiary Volcanics
Chuska Sandstone
Gallup Ss.
Mancos Shale
Dakota Sandstone
Morrison Formation
Entrada Sandstone
Chinle Formation
San Andres Fm.
Glorietta Ss.
U. Yeso Formation
Meseta Blanca Member
Abo Formation
Pennsylvanian

FIGURE 16   A Diagrammatic Cross Section Showing Geologic Relationships across the Southern Defiance Uplift and Chuska Mountains. Because the Defiance was an uplifted feature throughout Paleozoic time, many sedimentary rock units do not cross the uplift and others change formation names from the Black Mesa Basin to the west (W) across the basement structure into the San Juan Basin on the east (E).

Thus, the Defiance Uplift was a positive feature, standing above nearby seas throughout most of Paleozoic time (see figure 16).

Red mudstones and siltstones that occur between the Precambrian rocks and the overlying cliffs of DeChelly Sandstone have always been mapped as the Supai Formation, bringing Grand Canyon terminology into the Defiance region. These strata are most easily viewed at Hunters Point south of Window Rock, in Bonito Canyon west of Fort Defiance, and at the base of Spider Rock in upper Canyon de Chelly. This early assignment of names from Grand Canyon is somewhat misleading with regard to regional stratigraphic relationships, and when the Permian sections from nearby Utah and New Mexico are taken into consideration the interpretation is seen to be at least partly in error. It would be more appropriate to call this section the Abo Formation from New Mexico, as that undifferentiated red-bed sequence includes equivalents of the southern Utah Halgaito, Cedar Mesa and Organ Rock formations. Red beds directly beneath the DeChelly Sandstone are unquestionably those of the Organ Rock Shale, the equivalent of the Hermit Shale in Grand Canyon, not the Supai. Only the red ledgy cliffs in the lower red-bed section that support the Blue Canyon Dam west of Fort Defiance appear to be much like the Cedar Mesa Sandstone of southern Utah, the equivalent of the upper Supai Group, Esplanade Sandstone, in Marble and Grand canyons. In spite of these recently interpreted relationships, all geologic maps and guidebooks to the Defiance Uplift list the unit as Supai Formation. Perhaps only you and I know the real story.

Then there is the enigma of the DeChelly Sandstone that forms the magnificent red massive cliffs in Canyon de Chelly, where the name was derived, and the lower smooth cliffs at Hunters Point and in Bonito Canyon. This obviously windblown sand deposit thins rapidly onto the crest of the Defiance Uplift from more than 1,000 feet in the Black Mesa Basin to the west to 825 feet in Canyon de Chelly, finally becoming little more than 200 feet thick at Hunters Point. It thickens again to about 600 feet in the San Juan Basin east of the uplift, where it is called the Meseta Blanca Sandstone Member of the Yeso Formation and overlies the red Abo Formation. Not only was the Defiance Uplift still high during deposition of the DeChelly Sandstone, the structure remains the apex of confusion with regard to stratigraphic relationships and terminology. The problem results from the location of the Defiance Uplift; geologists from Arizona, Utah and New Mexico have studied the formation from different regional perspectives.

One long-time Arizona geologist, Wes Peirce, insists that the combined cliff-forming sandstone body at Hunters Point and in Bonito Canyon should be called DeChelly Sandstone, and he divides the section into several locally occurring named members. Indeed, the entire section has historically been mapped as DeChelly. However, after studying the New Mexico Permian, I concluded that the section at Hunters Point consists of the lower DeChelly Sandstone, a thin shaly slope-forming unit equivalent to the upper Yeso Formation (San Ysidro Member), and that the upper ledgy sandstone cliffs are the overlying Glorieta Sandstone of New Mexico, the stratigraphic equivalent to the Coconino Sandstone of Grand Canyon. After all, the upper Yeso is known to be thinning and changing to a shaly unit toward the north from the Zuni Mountains, and the upper flaggy sandstone cliffs, obviously deposited under different environmental circumstances (probably marine) at Hunters Point and Bonito Canyon appear identical to Glorieta exposures to the east. The latter interpretation simplifies the peculiar occurrences on the Defiance Uplift and eliminates the necessity for the many unwelcome complications in terminology. Since most folks really care very little, however, the terminology remains moot. The cliffs still appear on geologic maps as the DeChelly Sandstone.

MESOZOIC ROCKS

Sporadic stream deposits that form the basal member of the Chinle Formation, the Shinarump Conglomerate, rest directly above the DeChelly Sandstone throughout the Defiance Uplift north of about Hunters Point. Dark-brown mudstones of the Moenkopi Formation that generally intervene across Arizona and Utah are not present here on the uplift, again indicating that the Defiance was still in a positive mode in Triassic time. It is the Shinarump that is sufficiently resistant to erosion to support the high sandstone cliffs in Canyon de Chelly and to provide the surface of much of the Defiance Plateau. Upper members of the Late Triassic Chinle Formation are, as elsewhere, varicolored shaly beds that erode easily to form broad valleys. The valley that extends generally from Hunters Point northward to Lukachukai, known variously from south to north as Black Creek, Red, and Black Salt valleys, has been eroded from exposures of the upper members of the Chinle Formation, providing fine pasture and farm lands and a natural route for Navajo Highway 12. These valleys also divide the gently rounded Defiance Plateau from the Chuska Mountains that guard the skyline to the east.

The Wingate Sandstone, originally believed to be of Triassic age, is well developed in the Lukachukai Mountains, a northwestern extension of the Chuska Mountains near the village of Lukachukai. There, the Rock Point Member, originally included in the Wingate Sandstone, is now relegated to the Chinle Formation, and the Wingate Sandstone now consists of only the upper massive red sandstones of the formerly named Lukachukai Member. Recent work on fossil spores and pollen indicates that the Rock Point Member (of the Chinle Formation) is Late Triassic, while the Wingate Sandstone (Lukachukai Member) is Early Jurassic in age. The Wingate Sandstone is cut out toward the south by erosion at the J-2 unconformity in exposures along the base of the northern Chuska Mountains, also known as the Tunicha Mountains, and is not present south of a point near Crystal.

A thin red-bed sequence mapped as the Carmel Formation separates the Wingate from the next overlying red cliffs of the upper Entrada Sandstone, which continues southward through Window Rock to Lupton. There is no Navajo Sandstone on, or east of, the Defiance Uplift. The windblown deposits of the Navajo Sandstone ordinarily occur as a prominent light-colored highly crossbedded sandstone to the west between the Wingate and the Carmel formations. Above the Entrada, and perhaps really a light-colored facies of the Entrada, the Cow Springs Sandstone thickens rapidly and becomes a massive part of the sandstone facade southward through Window Rock and on to Lupton at the southward plunge of the Defiance Uplift. The famed natural arch of Window Rock is formed from the Cow Springs Sandstone. Where exposed, the Morrison Formation may be seen as rounded gray hills between the Cow Springs cliffs and the hogback of the Dakota Sandstone along the Defiance monocline from near Window Rock southward to Lupton. A red-bed sequence between the Cow Springs and the typical Morrison Formation was originally correlated with the Summerville Formation of the San Rafael Swell far to the northwest, but this unit is now referred to the Wanakah Formation.

The picturesque massive sandstone cliffs at Lupton have not clearly been identified to everyone's satisfaction. The rocks below the Entrada have been called "the beds at Lupton" by some USGS geologists, but others generally assign these to the Cow Springs Sandstone below and the Entrada Sandstone above a prominent bedding surface within the cliffs.

CHUSKA MOUNTAINS

Rising majestically from the colorful foothills of Jurassic sandstone cliffs, the Chuska Mountains dominate the skyline east of the Defiance Plateau. The range, along with the subranges of the Lukachukai and Tunicha mountains, consists of light-gray slopes and cliffs of the Chuska Sandstone and volcanic rocks of Tertiary age.

Light-gray, often appearing nearly white, lower slopes and cliffs of the Chuska Mountains are formed on the Chuska Sandstone. The formation is made up of a lower conglomerate and stream-deposited sandstone that lies unconformable above rocks ranging from the Chinle through the Dakota Sandstone. This lower unit reaches thicknesses of up to 250 feet. Above the basal stream deposits are highly cross-stratified undoubtedly windblown sandstones that form the bulk of the exposures. The formation averages about 1,000 feet in thickness overall and locally reaches 1,800 feet. Although its exact age is somewhat in doubt, it is usually assigned to the lower Tertiary because it has been cut by igneous dikes of Oligocene age and underlies a Pliocene volcanic sequence.

Dark-gray cliffs that cap the high ridges and summits of the Chuska Mountains consist of various kinds of volcanic rocks. Certainly tuffaceous rocks, formed of extruded ash and volcanic ejecta dominate, but intrusive sills and lava flows of potash-rich alkalic trachybasalts (called minette when in intrusive rocks) cap many of the topographic features in the range. Igneous sills have intruded rocks from the Mesozoic sandstone section to the Chuska Sandstone. Volcanic cones and plugs are common. The volcanic rocks are assigned a middle to late Tertiary age.

Diatremes, the vents of violently explosive gaseous eruptions from deep in the Earth's mantle, are scattered promiscuously about the Defiance Uplift as though there was no plan or pattern to their occurrence. The rocks may consist of breccias of metamorphic and granitic fragments and green high-temperature rock fragments brought up from the crust and mantle intermixed with fragments of sedimentary rocks, such as sandstone, having been carried to the surface by the eruption and then fallen back into the vent. Black Rock, just south of Fort Defiance, The Beast at the village of Navajo, Fluted Rock west of Sawmill, and Black Rock Butte between Canyon de Chelly and Canyon del Muerto are a few examples. Buell Park is another diatreme, not displayed as a sharp butte but rather as a circular valley. Here, kimberlite tuff, the rock from which diamonds are mined in other parts of the world, occurs in association with other breccias and minettes. Tiny garnets may be found concentrated in anthills near exposures of the kimberlite.

SCENIC DRIVES

The quickest and easiest way to reach the heart of the Defiance Uplift is to travel north from Gallup, New Mexico on U.S. Highway 666 for 8 miles. At *Yah-Ta-Hey*, the Navajo word for "welcome," keep left (west) onto New Mexico Highway 264 and drive 16 miles to Window Rock. The highway passes the massive open-pit McKinley Mine, operated by the Pittsburg and Midway Coal Mining Company, just east of the state line in New Mexico. Coal is strip-mined from about 5 zones in the Menefee and Crevasse Canyon formations of Late Cretaceous age. At the edge of the Navajo capital city, the Arizona and Navajo Indian Reservation border is appropriately marked by a sharp hogback of Dakota Sandstone, abruptly upturned along the Defiance monocline. The hogback also effectively separates rocks of Jurassic and older age to the west on the uplift, from Cretaceous-age rocks of the San Juan Basin to the east. Tribal Headquarters at Window Rock, named for a natural arch important in Navajo mythology, is nestled among steeply dipping exposures of Jurassic-age sandstone cliffs, namely the Entrada and Cow Springs sandstones.

WINDOW ROCK   Located at the Navajo capitol north of Gallup, New Mexico, this natural arch was formed by erosional processes from the Cow Springs Sandstone of Jurassic age.

NAVAJO ROUTE 12

The best and most scenic route for full enjoyment of the Defiance Uplift is to follow the length of Navajo Route 12 northward shadowing the Arizona–New Mexico border from Lupton through Window Rock to Round Rock. To reach Lupton, Arizona, travel west on Interstate Highway 40 for about 18 miles. The highway first crosses the western limb of the broad Gallup sag, a synclinal structure lying between the Zuni and Defiance uplifts, and then crosses the obvious Torrivio anticline, exposing rocks of Cretaceous age. From about the Arizona border west to Lupton, sedimentary strata begin their rise onto the southern plunging nose of the Defiance Uplift. First the Dakota Sandstone forms the rimrock above slopes of the Morrison Formation that give way abruptly downward into breathtaking cliffs of the Jurassic sandstones; the Cow Springs, Entrada and "beds at Lupton," in order, capture the attention of even the most skeptical. These magnificent sandstone cliffs end abruptly at the village of Lupton, where the Jurassic section lies on the varicolored mudstone slopes of the Triassic Chinle Formation.

Upon turning onto Navajo Route 12 at Lupton, it is soon noticeable that a sharp ridge of nearly vertical Shinarump Conglomerate Member of the Chinle Formation forms a natural barricade on the right (east), as the highway follows exposures along slopes of the Triassic Moenkopi Formation. The rounded slopes on the left (west) are weathered from the top of the Glorieta Sandstone (upper DeChelly) where it rolls over abruptly toward the east along the southern extension of the Defiance monocline. The highway generally follows this swale along the Chinle-Moenkopi shaly beds for several miles, crossing the Shinarump along a brush-covered ridge. There are occasional views of the colorful exposures of Jurassic sandstones to the east and the rollover on the Permian sandstones to the west, especially obvious at the mouth of Black Canyon.

The highway then enters a broad valley at Oak Springs. There are good views of the nearly vertical beds of the Jurassic sandstones forming hogbacks ahead to the right along the Defiance monocline, and there is a good view of the Permian section in cliffs in the distance ahead. Here, the tripartite nature of the DeChelly Sandstone is obvious where the section is exposed along a nearly west-east fault scarp. The smooth red lower cliffs beneath the skyline ahead are on the true DeChelly Sandstone, as here defined, with a prominent notch of upper Yeso Formation separating it from the overlying ledgy cliffs of the Glorieta Sandstone. A closer view of these relationships is spectacular at Hunters Point, where the upper Yeso is somewhat thinner and the section is dragged down sharply to the east along a drag fold adjacent to a northeasterly trending fault. Moenkopi and Chinle beds lie near vertical at the foot of the drag fold along the fault. The Moenkopi is not seen north of here along the Defiance Uplift, as the Shinarump Conglomerate Member of the Chinle Formation lies unconformably on the Glorieta or DeChelly Sandstones along and across the higher Defiance Plateau.

At the Junction of Navajo Route 12 and Arizona Highway 264, turn right and travel 2 miles to Window Rock, then left again on Navajo 12 toward Fort Defiance. Arizona Highway 264 crosses the crest of the Defiance Uplift along its route west to Ganado. A magnificent forest of ponderosa pine decorates the route above about 8,000 feet elevation.

Scenery along Navajo Route 12 from Window Rock north to Fort Defiance is relatively featureless except for the curvature of the Defiance Plateau rising westward and the Jurassic sandstones forming the valley's facade to the east. Then, just before reaching Fort Defiance, Black Rock, one of numerous diatremes on the Defiance Uplift, looms to the left, guarding southerly approaches to the historic village. Entering town, the highway leads westward and Navajo 12 turns abruptly north at the first intersection. Drive straight, turning right at the last moment, to view historic Fort Defiance, a compound of stone buildings nestled in a grove of trees that were young when Kit Carson gathered Navajo refugees together here in the winter of 1864 for the "Long Walk" to Fort Sumner and the first Navajo reservation.

The original fort was built at the mouth of the abrupt defile of Cañoncito Bonito ("Pretty Little Canyon" in Spanish) where access from the Defiance Plateau could be easily protected, according to the founder, Colonel Edward Vose Sumner. The outpost was established officially on 18 September 1851.

Frank McNitt stated:

> At Cañoncito Bonito, a place known to Navajos as *Tsehootsoh* (or Meadows in the Rocks), Colonel Sumner found what he regarded as an ideal site for a military post, a choice deplored by successive post commanders because on three sides the narrow canyon was hemmed in by almost vertical walls of rock and scrubby sand topped by a southward-tilting tableland that made the location not merely vulnerable but inviting to hostile attack. An attacking force could (and in time would) drop musket balls and arrows upon Sumner's fort below with about the ease that peas would be shelled into a pot upon a stove.
>
> Hell's Gate, as it soon became known to troops garrisoned here, officially was named Fort Defiance.
>
> Note 20: . . . Van Valkenburgh (1941) says the site 'was a favored Navajo rendezvous in the pre-American era. Medicine men here collected herbs known as *Le'eze'*, Horse Medicine, and the bubbling springs were shrines into which white shell and turquoise were thrown as payment or pleas for further blessings.'
> McNitt 1990, 195

Continuing with caution along the unpaved road with numerous sharp, blind curves, one enters Bonito Canyon to view magnificent exposures of the DeChelly Sandstone, actually thought to be the ledgy cliffs of the Glorieta Sandstone directly overlying the smooth pink cliffs of the true DeChelly Sandstone, in turn resting on the Organ Rock Shale (Supai Formation of local usage), all of Permian age. It is easy to visualize why this abrupt, precipitous defile was a special place for Navajos in the early days and one that had to be carefully guarded from surprise Indian attacks on the military fort.

The road to the left after Bonito Canyon leads to Blue Canyon Dam, with its abutments stapled into ledgy red sandstones and siltstones of the Supai Formation, more reminiscent of the red facies of the Cedar Mesa Sandstone of Utah. The road ahead leads to Quartzite Canyon, where the Supai lies unconformably on Precambrian quartzite.

Return to Navajo Route 12 in Fort Defiance and turn north (left). About 4.5 miles north of Fort Defiance, the Lukachukai Member of the Wingate Sandstone, forming rounded red cliffs below the Entrada Sandstone, makes its most southerly appearance at the base of the Jurassic sandstones on the right (east); distant skylines are upturned erosional edges of Dakota Sandstone. The Lukachukai Mountains are visible in the distance ahead. To the left (west) the Zilditloi Mountains ("Mountain with Hair on Top") are capped by columnar basalts of Tertiary age. At the village of Navajo, The Beast, a prominent diatreme composed of minette, rears its ugly head between housing developments and a large sawmill. A second, somewhat less obtrusive volcanic neck known as Outlet Neck lies west of Navajo Sawmill. Three miles beyond The Beast, Green Knobs lie just to the right of the highway. These are eroded from a lapilli tuff within a volcanic plug that has intruded into, and through, the Triassic Chinle Formation. Layers of the green tuff, which contain abundant fragments of exotic igneous rock types, dip inward in saucerlike fashion.

About 3 miles past the Green Knobs, the paved road to the right (New Mexico Highway 134) leads to Crystal and Sheep Springs via Narbona Pass. Formerly known as Washington Pass, the Navajo Nation Council officially renamed this important focal point in the Chuska Mountains Narbona Pass in 1992. This historic and infamous trail, leading from the plains of the San Juan Basin westward to the headwaters of the Canyon de Chelly drainage basin, is well worth a brief side excursion. Heading east, the highway traverses well-covered slopes of the obscured Chinle Formation, with strange-shaped knobs and goblins of Wingate Sandstone decorating the ridge to the left (north). Then well-weathered exposures of the Entrada and Cow Springs sandstones appear above Crystal, and an obscure outcrop of Dakota Sandstone at milepost 16 welcomes one to the inner circles of the southern Chuska Mountains.

From here the highway winds through lush forests of magnificent

NARBONA (WASHINGTON) PASS This photograph shows the eastern approach from the San Juan Basin. The pass was an easily guarded trail to Navajo strongholds on the Defiance Uplift through the Chuska Mountains and was the site of several battles during the 1800s. Rocks seen in and above the pass are of extrusive igneous origin of Tertiary age. The immediate area has been interpreted as a volcanic eruptive center.

ESCARPMENT FORMING THE WESTERN MARGIN OF THE SOUTHERN
LUKACHUKAI MOUNTAINS NEAR THE SETTLEMENT OF LUKACHUKAI
Lower slopes are in the Owl Rock Member beneath the lower cliffs of the
Lukachukai Member of the Wingate Sandstone. The Tertiary Chuska Sand-
stone here directly overlies the Morrison Formation of Late Jurassic age.

ponderosa pine that rise through scrub oak underbrush, past exposures
and roadcuts of the nearly white Chuska Sandstone of Tertiary age,
especially well developed near the Narbona Pass Campground. The
smooth sandstone bluffs are capped by black igneous rocks, often decorated with columnar jointing, that completely encapsulate the scene at
the pass. Exposures and roadcuts in and near the pass are in dark-gray
tuffaceous breccias. Above the pyroclastic deposits lie lava flows and a
rubble dome intruded by dikes and plugs. Everything suggests that this
topographic sag in the Chuska Mountains is associated with caldera
development—a volcanic center. The chaotic nature of these volcanic
features is perhaps best appreciated from the eastern approach to the
pass. From the topographic knoll formed at the top of a tree-covered
landslide terrain that masks the eastern foothills of the Chuska Range
for about 25 miles, the black jumble of past igneous holocaust strains
the imagination.

From Narbona Pass, looking back to the east onto the vast expansions of the nearly barren San Juan Basin and on to the west down the lush, ponderosa-studded grasslands of Crystal Creek, tributary to Coyote Wash and Canyon de Chelly, one cannot ignore the eerie sensation of the infamy and tragedies of history. The blackness of the scene is reminiscent of dark days passed. One can feel the bitter cold and bone-ravaging winds of a wintry day in early December 1804 when Lieutenant Colonel Antonio Narbona led his command of Spanish dragoons, perhaps through Narbona Pass (Washington Pass), to become hopelessly lost in the blizzard-ridden Chuskas, only to beat a humiliating retreat back to Laguna Pueblo. His mission had been to teach the Navajos who had been marauding Spanish settlements of north-central New Mexico to behave themselves. One Navajo warrior was killed, another was wounded, and a woman and two little girls were taken captive (McNitt 1990, 41).

He returned with a contingent of some 300 troops, accompanied by Opata Indian auxiliaries and led by Zuni guides. Clothed in broad-brimmed black hats and protective boot-length capes, carrying muskets at the ready and wielding glistening sabres, this motley militia probably again crossed what would be called Washington Pass to arrive at the rim of Canyon de Chelly on 17 January 1805. It encountered Navajo warriors that day, and on the following day heroically defeated "hundreds" of Navajo elders, women and children, hidden on a high ledge and in an adjoining cave several hundred feet above the floor of Canyon del Muerto, a northern tributary of Canyon de Chelly. Three hundred and fifty head of sheep and goats had been slaughtered; "Ninety warriors (apparently including old men) and twenty-five women and children had been killed. . . . He had come with more than ten thousand cartridges, but his guns had consumed almost all of them" (McNitt 1990, 43).

Following the ill-famed Narbona (Spanish) massacre of 1805, Narbona Pass was used frequently for encounters between Spanish, Mexican and American militia, both in peace and in bloodshed. Jose Antonio Vizcarra in 1823 led a large contingency of Mexican militia against the Navajos crossing Narbona Pass; his scattered attacks led to a short-lived treaty requiring that all Mexican captives be freed. Captain Blas de Hinojos and Don Juan Antonio Cabeza de Baca, with a Mexican force of more than a thousand men were ambushed and soundly defeated by Navajo warriors in Narbona Pass in 1835, their bodies left scattered amongst the pines. In 1849 Colonel John MaCrae Washington, Civil and Military Governor of New Mexico, crossed the pass,

CHAPTER TEN

originally named for him by Colonel Simpson, to negotiate a treaty between the United States and the Navajos at Canyon de Chelly. This was only after killing, scalping and beheading the important Navajo headman Narbona, an unfortunate act that led to years of warlike activities. Later in 1853, Henry Dodge established a temporary Navajo agency office near the pass.

How many other encounters were witnessed by the stately ponderosa is strictly conjecture; most reports complain of icy, bitter cold encounters with the declivity, a mood still evident today. Does that dark, icy mood reflect the ghosts of the dead of Narbona Pass?

From Narbona Pass, one may quickly travel down the massive landslide-laden eastern slope by modern highway to Sheep Springs, or return to Navajo Route 12 to continue the excursion of the Defiance Uplift.

Turning back north on Navajo 12, the view ahead is soon of Sonsela Buttes, capped by columnar-jointed lava flows resting on Chuska Sandstone over a base of Entrada Sandstone. The former home of Chee Dodge, important Navajo leader of the early 1900s, lies near the base of the right-hand butte. Then the highway provides a close-up view of beautiful Wheatfields Lake and campgrounds after passing White Cone, an outlier of Chuska Sandstone that lies unconformably on the Wingate Sandstone. Descending from the Chuska uplands, the highway passes the paved road to Chinle via the north rim of Canyon del Muerto, and then another to the right leading to the village of Lukachukai, with the colorful Lukachukai Mountains forming the scenic backdrop. Navajo Route 12 ends a few miles ahead, where it joins U.S. Highway 191 at Round Rock that goes on through Rock Point to join U.S. Highway 160 near Mexican Water.

A rough unpaved road leading east from Lukachukai crosses the Mesozoic sandstones, white Chuska Sandstone and various volcanic sills and flows of the Lukachukai Mountains. It passes the unique Dineh bi Keyah oil field, discovered in 1967 by Kerr-McGee Corporation and eventually returns to U.S. Highway 666 via Red Rock Valley, crossing the southern dike of Shiprock.

CANYON DE CHELLY

The name is mystical. It says nothing in Spanish, Navajo, French, or English, yet it says everything: deep canyons set in moody tones, quicksand, beauty, cornfields, peach orchards, the Navajo fortress feared by conquering soldiers, life, death.

The name "Canyon de Chelly" appears to be French; in Spanish it is pronounced—properly "de shay." However, this is the heart of Navajo Country and historians insist that the name is a corruption of the Navajo word *tsegi*, meaning "rock canyon," or "in a canyon" (Campbell Grant 1978, 3.) yet any translation of the sound *tsegi* into the sound "shay" makes little sense, phonetically or linguistically. I once thought that the name may have come from the French word *pêcher*, (pe shay) meaning peach tree, named for the once-rich peach orchards the early Navajos nourished in the canyon. (Kit Carson's crusades of 1864 cut down the orchards to starve out and demoralize the Navajos.) A Navajo interpreter at Canyon de Chelly National Monument explained that the name was derived from *Dineh Ch'ili'*, meaning "curly haired person," for a Navajo headman who once lived there. However, *The Navajo Language*, a dictionary by Young and Morgan, (1987, 732) suplies the most likely and little publicized answer: "Tseyi', Canyon de Chelly, Arizona (lit. inside the rock = canyon). The English name is quite apparently a French-like pronunciation of Spanish Chelly, which in turn is a phonological adaptation of Navajo tseyi'. Spanish substituted ch for ts and ll (=y) for Navajo y (=gh). Not knowing the meaning of the Navajo term, Spanish added Cañon de (Canyon of), on the premise that tseyi' was merely a placename. Thus, Canyon de Chelly could be construed as "Canyon of Canyon", after the fashion of Rio Grande River = Big River River)."

That settled, we can safely say that no visit to the Defiance Uplift would be complete without seeing Canyon de Chelly—"Canyon of Canyon."

PHYSIOGRAPHY

The general term Canyon de Chelly is often applied to the entire network of dendritic stream-cut canyons that drain the western flank of the Defiance Plateau. Indeed, the two main tributaries, Canyon de Chelly proper on the south, and Canyon del Muerto on the north, combine to become one, Chinle Wash, at the mouth of the canyon system; each of these in turn hosts dozens of minor tributaries, some given English names, others not.

Nearly vertical walls of the canyons result from the relatively resistant rock layers into which the ephemeral streams have eroded their channels. Softer, more easily eroded rocks above, the upper members of the Chinle Formation, have been almost completely stripped from the Defiance Plateau, a broad anticlinal fold. The basal member of the Chinle Formation, the Shinarump Conglomerate, however, is a hard, well-cemented conglomeratic sandstone that strongly resists lateral erosion, thus supporting the broad plateau and forming sturdy rimrock for the precipitous canyon system. As the Shinarump rests directly, although unconformably, above the DeChelly Sandstone (no ordinarily intervening Moenkopi Formation is here present), it has practically retarded lateral extension of the canyon walls and near-vertical cliffs result. Where the less resistant Organ Rock Shale (Supai Formation) is exposed beneath the DeChelly Sandstone, such as near Spider Rock, canyon widening has begun. Also near Spider Rock, the entire DeChelly Sandstone is exposed, here about 825 feet thick, and along with some underlying Supai and the overlying Shinarump Member, the canyon reaches its maximum depth of about 1,000 feet.

HISTORICAL SIGNIFICANCE

Although the scenic canyon network is in itself spectacular, its great importance from the earliest known times of human habitation in the American Southwest makes this a unique locale. It is not clear when the first Navajos drifted into Canyon de Chelly. Undoubtedly it was within the past 500 years, but they found not barren lands. Instead, the remains of an advanced culture in the form of dozens, perhaps hundreds, of deserted dwellings made of well-preserved masonry lined alcoves in the cliffs and the canyon floors alike. The Anasazi, the "ancient enemy," had preceded them, thriving here for a time and then for unknown reasons migrating on to better climes. Because Navajos respect the ghosts of the dead, they never frequented, or used, or damaged, the villages of stone. They must have wondered who these evasive pioneers had been, but quietly, without revenge, they established their own culture in the midst of this ancient abandoned niche. It is not difficult to understand why this was an ideal location. It was near summer pasturage and the hunting grounds of the high Defiance Plateau and the higher Chuska Mountains beyond, yet it provided a winter refuge and hiding place from one's predator enemies.

It was well known by the early Spanish conquistadors that Navajos lived here by the 15th century. Rumors that a great Navajo fortress existed in Canyon de Chelly caused considerable alarm and fearful concern to the first Spanish militia led into the canyon by Lieutenant Antonio Narbona in the winter of 1805. To its great relief, no fortress was found. Instead, the militia marched into the intricate canyon network in which hundreds, perhaps thousands, of Navajos could virtually disappear, only to reappear in the cliffs above, raining arrows and terror onto invaders below. This, in fact, was the nature of the feared "fortress."

It was in this setting that Narbona's troops found numerous old people, women and children—left hiding by Navajo warriors—on a high rocky ledge in Canyon del Muerto. A massacre followed, and several captives were taken at a place known to Navajos as "where two fell off," because a Navajo woman pulled a Spanish soldier from her hiding place 600 feet down the cliff to their deaths. However, this is not how the northern tributary canyon got its name. Colonel James S. Stevenson led an archeological search party here in 1882. Finding a pair of well-preserved mummies in the midden of an Anasazi dwelling in the upper part of the tributary, he named it Cañon de los Muertos ("canyon of the dead").

Although there were numerous abortive attempts by Spanish, Mexicans and New Mexicans alike, it took more than half a century for anyone to drive Diné from the canyons, and then it was through starvation imposed by Kit Carson's army during the 1864 crusade to put the Navajos onto a government reservation. When that escapade proved a failure to all concerned, and the resulting treaty was signed in 1868, the Navajos returned to their beloved red rock country.

Canyon de Chelly lies well within the so-called Treaty Reservation; however, in 1931 President Herbert Hoover announced the establishment of Canyon de Chelly National Monument within the Navajo Indian Reservation. Title 16, United States Code, Article 445 reads, in part:

> With the consent of the tribal council of the Navajo Tribe of Indians, the President of the United States is authorized to establish by presidential proclamation the Canyon De Chelly National Monument, within the Navajo Indian Reservation, Arizona, including the lands hereinafter described.
>
> All lands in Del Muerto, De Chelly, and Monument Canyons, in the canyons tributary thereto, and the lands within one-half mile of the rims of the said canyons . . . embracing about eighty-three thousand eight hundred and forty acres . . . in Arizona.

And in Article 445a:

> Nothing . . . shall be construed as in any way impairing the right, title, and interest of the Navajo Tribe of Indians which they now have and hold to all lands and minerals, including oil

and gas, and the surface use of such lands for agricultural, grazing, and other purposes . . . the said tribe is granted the preferential right . . . of furnishing riding animals for the use of visitors to the monument.

Thus, when visiting Canyon de Chelly National Monument, we are the guests of the Navajo Nation and must obtain Navajo guides for entry to the canyons, with the single exception of the White House Ruin trail on the south rim. We have the right, and the historical obligation, to view the wonders of these marvelous canyons from numerous viewpoints, both along the south rim of Canyon de Chelly and from the north rim of Canyon del Muerto. Here lies an unprecedented opportunity to look in from above on a culture little changed for decades. One can marvel at the lone spire of Spider Rock, trapped at the confluence of Chinle and Monument canyons, the domicile of legendary Spider Woman who taught *Diné* how to weave and make clothes. One can look in on Massacre Cave in upper Canyon del Muerto to wonder that scores of elderly women and children could occupy such a lofty perch, that Spanish militia would attain such precarious heights to destroy and capture the helpless. One can try to understand the Anasazi people who built and inhabited the remarkable ancient city still preserved in Mummy Cave and elsewhere in the intricate canyon network. This is the place to see true American history in one's mind's eye from a perspective not before realized by most of us.

CHAPTER ELEVEN

MONUMENT VALLEY

M onument Valley straddles the backbone of the Monument Upwarp, along the Mitten Butte syncline, only one of several folds
that chorus constitute an enormous uplifted asymmetrical, although locally wrinkled, anticlinal fold that extends from Kayenta, Arizona, for some 80 miles northward into Canyonlands National Park.
The eastern flank of the huge structure is the Comb Ridge monocline, a
sharp flexure formed where rocks of Paleozoic age drape over a basement fault zone. The opposing flank lies wherever one chooses, but generally it is defined by cliffs of the Mesozoic Wingate, Kayenta and
Navajo sandstones, known collectively as the Glen Canyon Group, that
border Canyonlands and Lake Powell to the west. As such, it is 30–40
miles wide. The Monument Upwarp formed in conjunction with other
classic monoclinal folds, such as the East Kaibab, Echo Cliffs, Water-
pocket, San Rafael and Defiance uplifts, as a consequence of strong,
easterly directed crustal stresses during the Laramide Orogeny in Late
Cretaceous to Early Tertiary time, some 65 million years ago. Even so,
the monoclinal flexures that bound these structures represent rejuvenation of basement fault zones, long since buried by sedimentary cover
rocks of Phanerozoic age.

The Monument Upwarp clearly distinguishes the downfolds of the
Henry basin on the west from the Paradox and Black Mesa basins on
the east. Because of the idiosyncracies of erosional processes, younger
sedimentary rocks preserved in the basins have been guarded from
removal by their lower, down-folded structural positions, while the
Monument Upwarp has been stripped unmercifully of these younger
rocks of Cretaceous and Tertiary age, exposing Paleozoic layered rocks
along the crest of the gigantic upfold. Of course, the folding process
caused by compression of the Earth's crust has naturally fractured the
brittle sedimentary cover along the axial trend, allowing erosion to
work at the broken strata from within to remove some blocks and preserve other more fortunate remnants. It is by coincidence of timing of

AN AERIAL VIEW OF MONUMENT VALLEY   This photograph looks north-ward toward Cedar Mesa on the far skyline. To the right is the barely visible Bears Ears. The rugged cliffs and spires are in the DeChelly Sandstone, some capped by the Hoskinnini Member of the Moenkopi Formation and the Shi-narump Member of the Chinle Formation, above lower slopes of the Organ Rock Shale. The Totem Pole and Yei Be Chei group of pinnacles may be seen in the right-middle of this photograph.

our human arrival on the scene, that the red beds of Permian age are seen in their last gasps of erosional survival in Monument Valley. Why one butte has withstood the onslaught of weathering while an adjacent valley has been stripped of its former cover by erosion is not known. Geomorphologists have a handy, highly technical term for this evasive process: "Differential erosion."

Highly fractured rocks at the crest of a large anticlinal fold, coupled with the coincidental depth of erosion to the level of red beds of Per-mian age, as seen at present time, have localized and created the scenic grandeur of Monument Valley. A most pleasant culmination of coincidental geologic features has occurred for our viewing pleasure.

KAYENTA

An automobile trip through Monument Valley may originate or terminate in either Kayenta, Arizona or Bluff, Utah, driving either north or south respectively on U.S. Highway 163. This discussion will begin at Bluff, first heading west, crossing the Comb Ridge monocline onto the Monument Upwarp to Mexican Hat, and then southward, tracing the crest of the anticlinal fold through the valley proper toward Kayenta at the southern plunging nose of the structure.

Bluff, Utah, founded in 1880 by Mormon pioneers who emigrated from various towns around Cedar City via a treacherous "short cut" now known as the Hole-in-the-Rock trail, lies on a pleasant terrace beside the San Juan River, rimmed by scenic cliffs of massive sandstone of Jurassic age. Although the country around Bluff has been frequented by Navajo people for hundreds of years, the legal Navajo Nation lies to the south of midstream of the river. The Reservation boundary runs from here to the confluence of the San Juan and the Colorado rivers, now inundated by Lake Powell.

The high, massive cliffs surrounding the valley have been named appropriately the Bluff Sandstone that also caps the Navajo Twins on the northeast edge of town. Beneath the impressive rimrock are chocolate brown thin layered siltstones of the Wanakah Formation, that overlie brightly red hues of the Entrada Sandstone, here a soft-weathering water-laid siltstone. Rarely exposed at the base of the cliffs is another brown siltstone, the Carmel Formation, forming the basal unit of the San Rafael Group, all of Jurassic age. To the south of Bluff, across the San Juan River on Casa del Eco Mesa (Spanish for Echo House Tableland), the Bluff Sandstone is capped by ledgy sandstones of the lower Morrison Formation (the Salt Wash Member); some geologists now consider the Bluff Sandstone as the basal member of the Morrison Formation. At shallow depths beneath Bluff, the Navajo Sandstone provides deliciously pure artesian well water to the community.

From Bluff to Butler Wash, 5 miles west of Bluff, the highway generally follows a bench formed near the top of the Navajo Sandstone (Jurassic), capped by terrace gravels derived from the San Juan Mountains of southwestern Colorado by heavy glacial runoff during Pleistocene time, fewer than 1 million years old. Two and a half miles west of the Bluff city limits is the turnoff to the left to Sand Island

Recreational Area where San Juan River trips originate (see chapter 12). In about another half-mile, the newly constructed highway leading across a bridge over the San Juan River toward Mexican Water is to the left. Then as the road dips downward toward Butler Wash, roadcuts reveal the true nature of the Carmel red beds overlying the top of the Navajo Sandstone.

After crossing Butler Wash, the highway climbs the steepening dip slope of the top of the Navajo Sandstone as it swoops up to meet Comb Ridge, the topographic expression of the Comb Ridge monocline. Cross-beds in the Navajo clearly testify to the windblown origin of the nearly white sandstone. At the deep roadcut at the crest of Comb Ridge, thinner bedding and smaller-scale cross-beds mark the crossing of the Kayenta Formation, guarding the massive pinkish facade of the Wingate Sandstone (Lower Jurassic) that oversees Comb Wash below. As road-cuts in the upper Chinle Formation (Upper Triassic) near the exit of the deeply carved portal unwisely undercut the Wingate cliffs, rockfalls are common and care should be exercised—it would not be wise to park or stand along the shoulder of the highway here.

From the base of the massive Wingate Sandstone to the floor of Comb Wash, the varicolored slopes of the Chinle Formation form the road footing on the right. Excellent views northward along the cox-combed flank of the monocline reveal the origin of the name "Comb Ridge." The sand-laden floor of Comb Wash masks the northward pin-chout here of the Permian DeChelly Sandstone and effectively masks any thin remnants of Moenkopi Formation that may be present. The highway crossing of Comb Wash is the practical northward limit of the DeChelly Sandstone, and the eastward limit—at least in outcrop—of the Moenkopi intertidal mudflat deposits of Early Triassic age.

Now U.S. Highway 163 begins its climb across the top of the Comb Ridge monocline. As the rocks are dipping steeply eastward and the highway is cutting down toward the west, rocks of Permian age are crossed quickly onto the older Pennsylvanian-age limestone. The first chocolate brown shales and siltstones are in the Organ Rock Shale that comprises the basal slopes of the buttes and mesas in Monument Valley to the south. Then, quickly, the roadcuts pass into pinkish siltstones, blebbed with white or light-gray gypsum, of the Cedar Mesa Sandstone. These represent the "lagoonal" facies of the usually massive sandstones of the Cedar Mesa. Looking to the right (north) near the crest of the Lime Ridge anticline a mile ahead, the change from massive sandstone

THE ORGAN ROCK MONOCLINE   This abrupt flexure in the sedimentary rocks of Triassic and Jurassic age across a deeply buried basement fault forms the western margin of the Monument Upwarp west of Monument Valley. Cliffs in this view across a Navajo corn field are in the Wingate Sandstone, overlying slopes of Chinle Formation. Cliffs on the far skyline consist of Kayenta Formation sandstones capping the Wingate cliffs at the edge of Skeleton Mesa.

to pinkish gypsiferous beds is apparent, occurring in a very short distance along the excellent exposures.

In the meantime, the highway again crosses reddish brown mudstones and siltstones. This time it is the Halgaito Shale, resting abruptly on a thin limestone at the top of the Honaker Trail Formation of the Hermosa Group of Late, but not latest, Pennsylvanian age. The Halgaito Shale, the lower unit in the Cutler Group, was originally considered to be of Early Permian age. However, with recent developments at the Permian type areas in the Ural Mountains of Russia and Kazakhstan, it is now seen to be of latest Pennsylvanian (latest Virgilian) age. The abrupt contact between the Halgaito red beds and the Honaker

THE VALLEY OF THE GODS Located north of Mexican Hat, this remarkable geological formation is sometimes called "the best-kept secret" in southern Utah. The region of photogenic cliffs and spires has been eroded from the Cedar Mesa Sandstone (cliffs) above Halgaito Shale slopes. The valley floor is formed at the top of the Pennsylvanian-age Hermosa Group.

Trail limestone, however, mark an erosional surface (disconformity) of regional significance.

The highway follows the uppermost limestone bed of the Honaker Trail Formation as it is folded sharply across the Lime Ridge anticline, the eastern fold of the huge Monument Upwarp, revealing in graphic detail the configuration at the surface of the anticline in all its glory. Just west of the crest of the fold, the limestone begins its descent westward into the northern Mexican Hat syncline, or downfold. Cedar Mesa, in the distance to the northwest, is the type area for the Cedar Mesa Sandstone that there reveals its typical cliff-forming characteristic. Sugarloaf Mountain, in the near distance to the left, is an erosional remnant of Halgaito Shale capped with some basal Cedar Mesa Sandstone. It is near the axis of a shallow syncline that separates Lime Ridge from

A DISTANT VIEW OF THE VALLEY OF THE GODS   This fairyland of desert scenery lies just beyond the Navajo Nation north of Mexican Hat, Utah. Cliffs and spires are in the Cedar Mesa Sandstone of Permian age.

Raplee anticlines that lie en echelon along the eastern Monument Upwarp. Monument Valley lies in the distance to the south, as the highway descends along the western flank of Lime Ridge anticline into the Mexican Hat syncline, around the northern plunging nose of Raplee anticline. As the highway enters the syncline a dirt road to the right leads into the Valley of the Gods, a valley eroded into the Halgaito Shale studded with picturesque pillars and mesas of Cedar Mesa Sandstone, much in the vogue of Monument Valley.

Mexican Hat, a butte of Halgaito Shale that appears much like a Mexican crouched in siesta-like repose with a balanced sombrero of Cedar Mesa Sandstone, lies in the Mexican Hat syncline to the left (east). Beyond is the San Juan River beneath the west-facing sharp fold of Raplee anticline. Ahead, just more than twenty miles from Bluff, is the junction with Utah Highway 261, leading northward onto Cedar

Mesa toward Natural Bridges National Monument. A road off Utah 261 to the left, less than a mile from the junction, leads to the Goosenecks of the San Juan overlook and State Park. There, one can view in stark clarity the meandering course of the river, now entrenched some 1,500 feet into the Monument Upwarp by downward cutting of the San Juan River during the past few million years.

Returning to U.S. Highway 163 and turning right to the village of Mexican Hat, Utah, the road travels through the Mexican Hat Oil Field. Miniature toylike oil pumps are reminders of the dashed hopes of E.L. Goodridge who drilled the discovery "gusher" here in 1908. The oil field, a freak synclinal accumulation where oil dribbles down-dip through porosity in shallow sandstones that occur above the water table, never produced oil in commercial quantities until a few barrels were sold to a sawmill at Blanding in the 1970s. According to the former owner of the San Juan Trading Post in Mexican Hat, the most often spoken words in the English language are: "How far is it to Kayenta?"

The highway crosses the San Juan River into the Navajo Indian Reservation on a third-generation variation of Goodridge Bridge at Mexican Hat village and immediately begins to climb out of the Mexican Hat syncline onto the east flank of the Cedar Mesa anticline that forms the backbone of the Monument Upwarp. A mile and a half south of Goodridge Bridge, Alhambra Rock, an erosional remnant of a basaltic (minette) volcanic dike rises above the crest of the anticline ahead. Then a road to the left leads to an abandoned uranium concentrating mill and town site that processed ore from the Vanadium Corporation of America Monument Valley No. 2 mine that was located on the east side of Monument Valley, producing more than 5 million pounds of U_3O_8 during the years 1942–69. Views ahead of Monument Valley pique the imagination. Long, straight stretches of the highway render the approach endless.

At Redlands Overlook, Monument Valley finally seems real. From left to right, visible are Castle (Cathedral) Rock, King on His Throne, Saddle Rock, Monument Pass, and Eagle Mesa beyond Setting Hen. The valley floor consists of thin sandstone and limestone beds in red siltstones of the Cedar Mesa Sandstone, here trying to change into the lagoonal facies of the Four Corners region to the east. These beds have been mistaken for the upper Honaker Trail Formation by some recent roadlog entrepreneurs. Lower reddish brown slopes in the buttes and pinnacles ahead consist of the Organ Rock Shale, overlain by the massive cliff-forming brown walls of the DeChelly Sandstone, all of Per-

MAGNIFICENT CROSS-BEDDING IN THE DECHELLY SANDSTONE (PER-MIAN) This cross-bedding is well exposed in Canyon de Chelly, here seen along the White House Ruin Trail within the national monument.

mian age. Capping the spires are thin slope-forming remnants of the Lower Triassic Moenkopi Formation, capped on most "monuments" by isolated bits of the Shinarump Member of the Chinle Formation of Late Triassic age.

The road from here to Monument Pass wanders through exposures of the lagoonal Cedar Mesa Sandstone that tops out at the Pass. All around us at Monument Pass are buttes consisting of basal Organ Rock Shale slopes, DeChelly Sandstone cliffs, with thin Moenkopi and Shinarump caps. From left to right are Big Indian, Sentinel Mesa, Gray Whiskers Butte, Mitchell Butte, Rock Door Mesa, Eagle Mesa and Setting Hen.

In fewer than 3 miles, barren lands of magnificent red beauty give way to civilization. A road to the right leads to Monument Valley tribal boarding school, Goulding's Trading Post and lodge in Rock Door Gap, a quick food and gasoline refreshment post, and beyond, commercial campgrounds and *Oljeto* ("Place of moonlight water" in Navajo) Trading Post.

To the left of the same intersection is a road to Monument Valley Navajo Tribal Park. After stopping to deposit nominal Tribal Park entrance fees, one is finally treated to the real Monument Valley of Hollywood fame, replete with the East and West Mitten Buttes—international trademarks. A driving tour of about 14 miles through the 30,000-acre Tribal Park will leave indelible memories and dozens of color photographs of Monument Valley for posterity. Encountered en route are Sentinel Mesa, the Mittens, Merrick Butte, Mitchell Mesa, Elephant and Camel buttes, Spear Head Mesa, Rain God Mesa, and Thunder Bird Mesa, all diminished in splendor by a magnificent view of the Totem Pole and Yei Bichai figures at journey's end.

Unlike what may be found in some published roadlogs of the area, all topographic features in Monument Valley consist of lower slopes of Permian Organ Rock Shale and massive cliffs of DeChelly Sandstone, with thin remnants of Triassic Moenkopi and Shinarump capping some points.

The Utah-Arizona state line is crossed just south of the Monument Valley highway intersection. Beyond this point the road gradually leaves Monument Valley behind. In less than a mile, the next landmark, Agathla Peak (or El Capitan), captures the view. It is another diatreme, or gaseous eruptive volcanic vent of Tertiary age like Shiprock, which stands sentinel to Monument Valley's southern approach. The Organ Rock monocline is visible to the right (west), capped by Hoskinnini Mesa and the Wingate Sandstone; *Hoskinnini* (meaning "the generous one") was a Navajo headman who led his followers into this primitive, stark region to avoid capture by Kit Carson in 1864.

The DeChelly Sandstone slowly works its way down to road level, as the underlying Organ Rock Shale descends into the subsurface southward along the plunge of the Monument Upwarp; then the DeChelly Sandstone disappears into the depths in turn. Beyond Boot Mesa, rocks of Triassic and then Jurassic age dominate the view. The various named members of the Chinle Formation are well expressed below Owl Rock to the right (west) of Agathla. The basal Shinarump Member hosted uranium deposits mined in this vicinity during the boom days. Klink (1973) wrote: "Owl Rock is absurd. It shouldn't be there at all. Passive

THE TOTEM POLE *(right)* and the YEI BE CHEI DANCERS *(left)*   These two spectacular formations mark the southern limits of Monument Valley. Rocks seen in the cliffs and spires are in the DeChelly Sandstone, rising above lower slopes of Organ Shale. Both are Permian in age.

and unconcerned it sits there on the plateau across from Agathla, glaring down at each passer-by, massive and silent in this land of the monuments, where massiveness and silence stand supreme."

Rounding a bend to the left, Agathla looms ahead. Named El Capitan by Kit Carson, the Tertiary-age volcanic vent, technically a diatreme (violent blowout vent), rises 1,225 feet above the surrounding countryside, creating a landmark supreme. The ominous gray goliath is the site of annual sheep-shearing gatherings by Navajos, thus the name Agathla: "the place of the scraping of the hides."

After passing Agathla, the view ahead is dominated by Black Mesa, a structural basin in which rocks as young as the Late Cretaceous

Mesaverde Group, namely the Yale Point Sandstone, cap the skyline. As the highway follows the southward plunging structural nose of the Monument Upwarp toward the village of Kayenta, the Chinle Formation, followed in turn by the younger Wingate Sandstone, Kayenta Formation (here at its type section), and Navajo Sandstone, dives into the subsurface, disappearing beneath the ever-increasing dominance of Black Mesa ahead.

Kayenta, Arizona—"The farthest point from anything" in the United States—was first settled by whites when John and Louisa Wetherill opened their Tyende Trading Post here in 1910. The name "Kayenta," pronounced something like a quick and sloppy "Kanta" by locals, is a pronunciational corruption of the Navajo word *Tyende*, meaning "where the animals bog down." The Wetherills were instrumental in the discovery of Mesa Verde's Anasazi Indian ruins and later discovered Keet Seel, Inscription House and Betatakin, now Navajo National Monument; John was the first white man to see Rainbow Bridge, now another National Monument on Lake Powell.

CHAPTER TWELVE

SAN JUAN RIVER

I t is the San Juan River that provides much of the northern limit to Navajo Country. Indeed, downstream from the mouth of Recapture Creek, just east of Bluff, Utah, the center of the San Juan River as it occurred in 1884 is the legal boundary of the Navajo Indian Reservation. Because a number of Navajos have traditionally farmed and grazed lands north of the river, there are several tracts of allotted lands there, and the Reservation boundary was extended several miles to the north in 1905 and 1933, and again in 1958 in a trade for land at the site of Glen Canyon Dam and Page, Arizona.

The river forms a natural barrier to travel, and as such it is the traditional marker between Navajo Country and Ute territory. Although it can be crossed easily on foot or horseback for much of the year above the mouth of Chinle Wash, most traditional Navajo people have a healthy respect for the power of the river and avoid it. During normal high water, which occurs in late May and early June, the San Juan can be a raging torrent. However, it was known to dry up completely in late summer in historic times. Downstream from the mouth of Chinle Wash, the river has carved formidable canyons, and normal access to the river is difficult to impossible.

Headwaters for the San Juan River are in the high San Juan Mountains near Wolf Creek Pass, east of Pagosa Springs, Colorado, with the Animas River draining much of the western San Juans near Silverton and Durango, Colorado, forming a major tributary. These mountains that attain elevations greater than 14,000 feet provide the first obstacle to storms coming out of the southwest, and during most years receive the greatest snow packs in the Southern Rocky Mountains. Below the mouth of the Animas, near Aztec, New Mexico, the San Juan becomes docile and flows promiscuously across the northern San Juan Basin and through the oil fields of southeastern Utah before entering the spectacular canyon country west of the Comb Ridge monocline. Because the river is quiet and the scenery is relatively uneventful, most river

THE SAN JUAN RIVER   This river has carved its course across Comb Ridge *(right)* and enters its upper canyon through the Lime Ridge anticline *(left)*. Rocks seen in this view from atop the Mule Ear diatreme range from the Navajo Sandstone (Jurassic) on the right down to the Hermosa Group (Pennsylvanian) in the prominent upfold of the eastern Monument Upwarp on the left. The Abajo Mountains, a Tertiary-age laccolithic intrusive igneous range, forms the distant skyline to the left of the desert "male rain" storm. (Photograph by Gene Stevenson.)

trips begin at the last point of easy access at Sand Island Recreational Area where the U.S. Bureau of Land Management maintains a boat-launching ramp about 4 miles west of Bluff, Utah.

The San Juan is a relatively serene river, having no major rapids with which to contend. Consequently, it is good training for the neophyte "river rat." Nevertheless, no major river of this proportion should be taken lightly. The gradient of the San Juan River is steep, and the flow is fast as most rivers go, especially within the deep canyons downstream. The few rapids are relatively small but are very rocky, especially in low-water stages, and are to be treated with considerable caution. When in doubt, any good boatman stops at the head of an unknown rapid to scout its momentary condition and to plot a safe course for a run that could otherwise end in disaster.

It is foolhardy to enter the canyons with anything less than the best in river-running equipment. Inflatable rafts built for the purpose are highly recommended, and dependable life preservers must be worn at all times. Everything one needs for camping, cooking, boat repairs, and first aid, and drinking water must be carried into the canyons, and all waste, including solid human waste, must be carried out. Rafting permits are required by the United States Bureau of Land Management and the Glen Canyon National Recreation Area in the lower canyon, whether the trip is private or conducted by professional outfitters; contact the BLM in Monticello, Utah for details. Above all, be thoroughly prepared in advance for any or all possible difficulties that may occur along a wilderness river, and you will have a safe and most enjoyable experience.

DOWN THE RIVER

The campground and boat ramp at Sand Island are guarded by massive, light-colored cliffs of the Jurassic Navajo Sandstone, which define the course of the river for the first 7 miles. Very thick cross-bedding sets attest to the eolian (windblown) origin of the extensive desert deposits. Alcoves in cliffs of Navajo Sandstone provide shelter for Anasazi ruins at River House and in Chinle Wash, as well as for the mud nests of thousands of swallows. Above the Navajo cliffs rise slopes eroded from the San Rafael Group, capped by the brownish cliffs of the Bluff Sandstone, all of Jurassic age. Formations of the San Rafael Group are difficult to differentiate here, due to the seemingly unusual presence of a waterlaid siltstone facies of the middle Entrada Sandstone, which is distinguishable from the Wanakah (Summerville) red beds above and the Carmel red beds below only by its bright orange color (see figure 17).

The river is slow-running for the first 9 miles, and its bifurcating channels meander freely, sometimes shifting almost daily. A boatman skilled at reading the river follows the trace of bubbles and Indian soap suds to avoid hidden underwater sand bars, but even the best are sometimes fooled and end up pushing the boat across shallow reaches. Terrace gravels left stranded well above present-day river level remind us of wilder times when heavy glacial outwash deposited well-rounded cobbles and boulders derived from the San Juan Mountains along the river's course some 15,000–1 million years ago.

Points of interest along the river are shown in the waterproof river guide published by Cañon Publishers Ltd., including a spectacular panel of rock art at about Mile 4.4 and Anasazi ruins at Mile 5.7. Just below

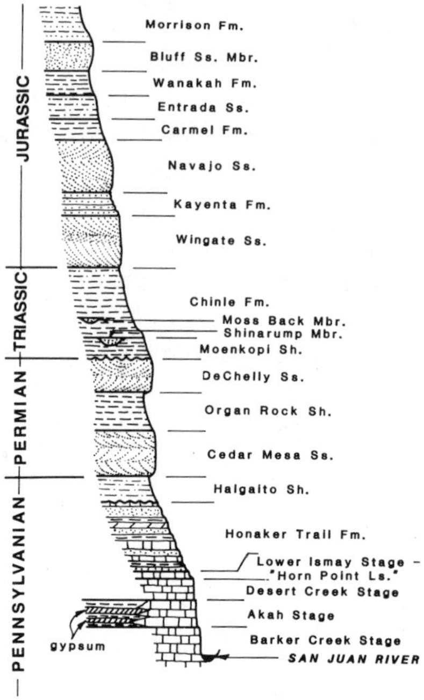
Morrison Fm.
Bluff Ss. Mbr.
Wanakah Fm.
Entrada Ss.
Carmel Fm.
Navajo Ss.
Kayenta Fm.
Wingate Ss.
Chinle Fm.
Moss Back Mbr.
Shinarump Mbr.
Moenkopi Sh.
DeChelly Ss.
Organ Rock Sh.
Cedar Mesa Ss.
Halgaito Sh.
Honaker Trail Fm.
Lower Ismay Stage - "Horn Point Ls."
Desert Creek Stage
Akah Stage
Barker Creek Stage
SAN JUAN RIVER
gypsum
JURASSIC
TRIASSIC
PERMIAN
PENNSYLVANIAN

FIGURE 17 Stratigraphic Column Showing Named Rock Units and Their Outcrop Appearances As Seen in the Four Corners Area and along the San Juan River Downstream from Bluff, Utah.

River House, the rock layers begin to rise, first rather gently, but then with a vengeance as the river approaches the Comb Ridge monocline. The layered rocks arch upward to the west along the sharp fold, first exposing stream deposits of the Kayenta Formation, then the reddish cliffs of Wingate Sandstone at Comb Ridge proper. Varicolored slopes of Chinle Formation provide a marker in the section beneath the Wingate cliffs, but the underlying thin brown mudstones of the Moenkopi Formation, are inconspicuous. A red cliff marks the near-vertical bed of DeChelly Sandstone that extends as straight as a string southward into the gray rubble pile of the Mule Ear diatreme, another neck of a violently eruptive, gaseous volcano that broke through the crust along the major basement fault underlying the monocline. A smattering of reddish brown Organ Rock Shale is visible just beneath (to the west of) the DeChelly exposure. Looking north, pink and white beds of the lagoonal facies of the Cedar Mesa Sandstone are conspicuous, with dark-brown flatirons of Halgaito Shale forming the scalloped cliffs along Lime Ridge anticline ahead to the west. The gray skyline and upper slopes consist of limestone beds in the upper part of the Honaker

THE UPPER CANYON OF THE SAN JUAN RIVER Seen here from the top of the Mule Ear diatreme, the canyon has been carved across the upfold of the Lime Ridge anticline on the eastern margin of the Monument Upwarp, exposing sedimentary rocks of the Hermosa Group of Pennsylvanian age. (Photograph by Gene Stevenson.)

Trail Formation of Pennsylvanian age. All of this section of Triassic through Permian rocks is crossed by the river in a half-mile.

Then at Mile 9.3, the river passes its closest approach to the Mule Ear diatreme, and plunges headlong into the upper canyon of the San Juan River, with cliffs composed of the Honaker Trail Formation. A 2-mile round trip to the top of the diatreme, not to be considered lightly in hot weather, provides a magnificent view of the San Juan Valley and Comb Ridge monocline and a chance to view tiny garnets from the Earth's mantle, having been discarded by ants in their favorite anthills near the summit. (Navajo Nation law prohibits collecting gemstones or other "rockhound" materials from within Reservation boundaries; all geological investigations must be conducted by permit only.)

Back on the river, the scenery changes abruptly from open land with low cliffs to a deep, awesome canyon of alternating cliffs and slopes of marine limestones and shales that reach to the skyline. Most of the beds in the canyon walls contain fossils of organisms that lived in the sea some 300 million years ago; brachiopods, corals, crinoids and bryozoa dominate the fossil biota, but the remains of nearly every variety of late Paleozoic marine animal may be found here, and many trails of burrowing wormlike critters are present as well. Many of the fossils consist of red chert (jasper), a secondary iron-rich form of silica that commonly replaces limestone particles, such as fossils.

A tributary canyon on the right at Mile 12.4 is the outlet of an abandoned meander, incised by the river during its initial canyon-cutting stage, later cut off by erosion and left stranded by the canyon-deepening process. One can walk around the loop from right to left following the ancient course of the river, the inflow and outflow canyon now being the same wash that empties into the river, perched about 50 feet up the canyon wall. In a sharp meander such as this one, the river's erosive power is directed toward the outside of a bend in the river, where erosion is greatest. As the river rounds the "gooseneck," erosion is thus directed at the thinnest separation at either end of the loop, both as it enters and then leaves the loop, and the narrow neck is eventually cut off, shortening the course of the river and steepening its gradient. Mark Twain noted that the course of the Mississippi River had been shortened by this process by several tens of miles, suggesting that it once must have jutted out over the Gulf of Mexico like a fishing pole.

Two miles below the abandoned meander, a sickly yellow bed of cherty dolomite rises to river level. This marker bed, known locally as "Old Yeller," is a weathered and silicified version of a bed of black shale that is found throughout the Paradox Basin to the north, where it was the source bed of voluminous deposits of oil now being produced in the Four Corners region. It occurs at the top of a cliff-forming limestone bed, known to geologists as the Lower Ismay oil zone. As the Lower Ismay rises to river level, it is first seen as a rolling surface that appears as small anticlinal-synclinal folds. When the lower flat-lying surface of the bed is seen above Eight-Foot Rapid, it becomes apparent that the rolling surface is the top of a layer that consists of numerous limestone buildups, or bioherms. The rock consists of local deposits of fossil calcareous algae that lived on shallow banks and accumulated, as they died, into piles of calcareous fragments that now resemble fossilized corn flakes, or smashed potato chips. Consequently, the spaces between the broken fragments of algae have been preserved as pores in the rocks, and these contain oil in the subsurface to the east, especially well developed in the Aneth Oil Field and other smaller fields in the Four Corners region. The layer is especially accessible near the head of Eight-Foot Rapid, where one can see a rare example of a depleted oil field exposed at the surface.

The river has carved its course deeper into the section, exposing a massive cliff of the Desert Creek oil zone from the head of Eight-Foot Rapid through The Narrows, emerging into black shale and gypsum slopes of the still older Akah oil zone in Soda Basin on the crest of Raplee anticline. It was here, in 1928, that Utah Southern Oil Company drilled a dry hole to a depth of about 1,900 feet in search of oil believed to occur in the crests of such upfolds (anticlines). The venture caused the laborious construction of a primitive, now impassable road that leads from near Mexican Hat rock into Soda Basin, visible along the right bank of the river below Ledge Rapid. At that point, the river emerges into open country for a brief respite through a portal cutting the west flank of Raplee anticline, another monoclinal fold developed where sedimentary rocks have been draped over a large basement fault block.

Leaving the gaping chasm of the upper canyon, the San Juan River lazily wanders through red rocks of the Halgaito Shale in Mexican Hat syncline, a downfold in the layered rocks made quite conspicuous by desert erosion. E.L. Goodridge noticed oil seeps in the canyon walls at what is now the Mexican Hat bridge on a river trip in search of gold during 1879–80, and in 1907 returned to drill a "gusher" near the present town site of Mexican Hat, opening the Mexican Hat Oil Field. The oil occurs in a rare synclinal deposit where oil dribbles down the bedding into the bottom of the fold where there is no groundwater to buoy up the lighter substance (Wengerd 1955). The second phase of drilling was initiated by a local bar owner in search of daytime excitement, and with sound developmental assistance from Durango geologist Bob Lauth, the field now produces oil in commercial quantities.

GOOSENECKS OF THE SAN JUAN

Passing under the bridge at Mexican Hat, the river again descends into the depths of deep canyon walls as it crosses the broad backbone of the Monument Upwarp, perhaps better known as the Cedar Mesa anticline. As one slowly floats between ever-rising walls of alternating limestone and gray shale of Pennsylvanian age, it gradually becomes apparent that there is no going back, that telephones and honking automobiles no longer exist, that one's well-being for several days to come is of one's own creation.

Before many miles into the lower canyon, the river begins to flow in tight, entrenched meanders. It sometimes flows for several miles to gain a distance of a few feet on its westward course. First, there is Menden-

GOOSENECKS OF THE SAN JUAN RIVER These deeply entrenched meanders have been incised more than 1,000 feet into the Pennsylvanian-age Hermosa Group on the Monument Upwarp. A two-lane paved road leading to Goosenecks Utah State Park, seen in the middle left of this aerial view, provides scale.

hall Loop, then The Tabernacle, followed in short order by The Goosenecks, as the canyon deepens to some 1,500 feet nearing the crest of the upwarp. Between Mexican Hat and Honaker Trail in the deepest part of the canyon near the structural crest of the giant fold of rock, the boat must travel 17 miles by river to gain only 10 miles geographically. Then the layered rocks begin their slow, ever so gradual descent from the axis of the Monument Upwarp westward into the Henry Basin. Then between Miles 47 and 51 the river again forms deeply entrenched meanders that are nearly identical to those in The Goosenecks. Until Honaker Trail, the river gradually cuts down in the stratigraphic section, traversing older and older rocks as it progresses. The process is reversed as the river cuts its canyon down the western flank of the upwarp west of Honaker Trail, and younger rocks progressively dip beneath river level and disappear from view.

Honaker Trail was built to supply prospectors that swarmed into the canyon during the gold rush of 1892. Construction was not begun until January 1893 and because the first horse down the trail fell to its death while attempting to descend from Horn Point midway down the canyon walls, the trail was never successful as a supply line. The gold rush ended as quickly as it had begun, and the trail fell into almost immediate disuse. A hike up the trail from Mile 44.3 can be extremely arduous and hot in summer; drinking water must be carried. However, the view of the canyon from Horn Point can make the climb well worthwhile.

About midway down from the skyline at the top of the Lower Ismay oil zone, immediately below the distinctive exposure of "Old Yeller," the ledgy slope of the upper canyon walls breaks into massive limestone cliffs that form the inner gorge. Although these cliffs are continuations of layers seen in the upper canyon, the rocks have changed beneath the Mexican Hat syncline. The upper, thinner cliff of Lower Ismay is no longer mounded and contains little algal material. The next, lower but thicker cliff of Desert Creek limestone is more massive than seen in Raplee anticline. Slope-forming dark-gray shale and bedded gypsum in the Akah cycle of the upper canyon are now seen to be a massive, thick, cliff-forming limestone; the change marks the edge of the Paradox evaporite basin. Here also the next underlying zone, the Barker Creek, forms insurmountable cliffs at river's edge. Both the Akah and Barker Creek layers now contain small, isolated mounds (bioherms) that appear to interconnect from canyon wall to opposite wall in northwesterly oriented trends. They are commonly seen along the northwesterly oriented course of the river that overlies an extension of the Defiance anticline of the Chuska Mountains, here projected northwestward from Mexican Hat to about Slickhorn Gulch. Certainly, basement faulting has indirectly controlled the shape of the basin margin and belt of conditions favorable to mound development in Middle Pennsylvanian time.

Rapids first appear in the lower canyon at Mile 52.2. The first, Ross Rapid, was named for Kenny Ross, a pioneer commercial river runner on the San Juan who founded Wild Rivers Expeditions in the 1940s. It is merely a quickened flow around a boulder fan on the right

THE LOWER CANYON OF THE SAN JUAN RIVER NEAR JOHNS CANYON
Rocks of the Pennsylvanian Hermosa Group form the inner canyon walls, with Muley Point, capped by the Cedar Mesa Sandstone of Permian age, the high point on the center skyline.

with a large boulder midstream in low water; in flows exceeding 10,000 cfs (cubic feet per second), it becomes a notable rapid with a lively ride. Next is Johns Canyon, always a rocky riffle skirting another boulder fan at the mouth of the canyon on the right. Then Government Rapid at Mile 63.4 is a nearly impassable rock garden in low water of about 500 cfs, but becomes a most lively run in flows of 1,000–10,000 cfs, only to wash out in higher flows. It always pays to stop and look at Government Rapid to avoid painful surprises.

Just around the corner is Slickhorn Rapid (Mile 66.3) at the mouth of Slickhorn Gulch, a long, beautiful tributary from the north that contains multitudes of little waterfalls and plunge pools, some designed for high diving and swimming when the streamlet is flowing and thus kept clean. A leisurely walk up Slickhorn is a must. The side canyon is narrow and deep, the atmosphere is one of intense solitude, yet the environment is that of a remote spring-fed wash bathed in sentiments of the desert Southwest. Three (four in low water) campsites beckon the river traveller, but all can lose their native charm when overcrowded by last-night-on-the-river campers. The rapid is a fast maneuver between large boulders, scattered randomly midstream off the terminus of the Slickhorn Gulch boulder fan.

Fluctuating levels of Lake Powell bring retarded river flow and sediment buildups into varying juxtaposition with the foot of Slickhorn Rapid. When the reservoir is full, sluggish water flow is noticeable a mile below the rapid, but when the lake is low, current is welcome to below Clay Hills Crossing at Mile 83.5. With low lake levels, a waterfall has developed between Clay Hills Crossing and Paiute Farms, making exit impossible below Clay Hills. Rock layers continue to plunge below river level between Slickhorn Gulch and Clay Hills. First, the uppermost of the Pennsylvanian limestones disappears into the subsurface at Mile 72, then the Halgaito red beds disappear near Buckhorn Canyon at Mile 75.2, leaving only low white cliffs of the Cedar Mesa Sandstone to guard the course of the river/lake on to Clay Hills. Varicolored slopes and cliffs west of Clay Hills are exposures of Mesozoic strata up to the Navajo Sandstone, similar in appearance to those seen in Comb Wash upstream.

The canyons of the San Juan River are remarkable showcases for rocks of the northwestern limits of Navajo Country and provide exceptional views of exhumed geologic structures. They were never domiciles to many Anasazi or Navajo families, but the feeling of ancient solitude and nearness to nature's beauty prevails as nowhere else.

GEOLOGY IN THE GRAND CANYON

Since the first trip through Grand Canyon in 1869 by John Wesley Powell, a self-taught geologist, the canyon has been considered the greatest natural geological laboratory on the planet. The Colorado River has cut a mile-deep chasm through the Earth's crust for 280 miles, and in its relatively arid climate the exposures of eons of Earth history are seen in magnificent detail. As Powell put it, "All about me are interesting geological records. The book is open, and I can read as I run."

The Colorado River rises in the high country of the Colorado Rocky Mountains at the foot of Longs Peak in Rocky Mountain National Park and is doubled in volume and muscle at its confluence with the Green River in the heart of Canyonlands National Park in Utah. The mighty river has been crippled by the man-made Glen Canyon Dam near the terminus of Glen Canyon but plunges headlong into Marble Canyon at Lees Ferry, Arizona. It must be at this point that travellers into Grand Canyon embark, as this is the only vehicular contact with the river between Hite, Utah and Hoover Dam, a distance of 500 river miles. (Primitive access is possible from Peach Springs via Diamond Creek at Mile 226 below Lees Ferry.) It is also exactly here that the river descends abruptly into the depths of canyons cut into rocks of Paleozoic age, the ancient ones.

It is common knowledge that rivers cut canyons with the relentless grinding power of silt and sand scraping at their bottoms but no one knows why the Colorado River chose this site to desecrate the Colorado Plateau. There have, of course, been many educated guesses.

There is little doubt that the great rivers of the Colorado River system achieved courses similar to today's by middle Tertiary time, some 30 million years ago. Then, the landscape we know today was buried under at least 5,000 feet of sedimentary rock that has since been washed away by erosion to some distant settling pond. Most went to the Pacific Ocean via the Gulf of California. Then, too, the land was flat, lying near sea level, with little gradient to encourage the rivers, and they meandered aimlessly. Soon a day came when the land began to rise. The entire western North American continent was gradually and very slowly uplifted, and the rivers began to cut downward into the land. Soon the rivers had entrapped themselves in valleys and canyons from which they could not escape, and as the land continued to rise, the rivers became entrenched in their own deep canyons. As older and harder rocks were encountered, with ever-increasing powers the rivers were forced to keep cutting or become impounded to form great lakes. Even then the increasing drainage gradients caused the waters to over-flow the hard-rock dams and continue on towards the sea. In this way, canyons were carved deep into the hardest rocks on Earth, and canyons like the Grand were made eminent. Rocks formed as shallow water sed-iments more than 250 million years ago were elevated to heights of more than a mile above present-day sea level and were exposed to view by canyon-cutting processes to the primitive horses and camels that grazed the countryside.

Erosion tries with all its might to level the land. The process is grad-ual. Starting at the highest levels, powers of erosion eventually plane down the land toward a terminal base level that is the surface of the sea. Where rocks are downfolded into broad basins, younger rocks are preserved longer, but where the rocks have been uplifted onto high folds, the younger rocks are readily stripped from the land, exposing the most ancient rocks on the structurally highest uplifts. It is just this situa-tion that begins at the present site of Lees Ferry to allow the Colorado River to expose the most ancient rocks of the northwestern margin of Navajo Country in the depths of Grand Canyon.

At Lees Ferry a long, asymmetrical upfold—the Echo Cliffs mono-cline—brings older rocks up to the surface of the land from the Black Mesa Basin to the east to form the Marble Platform. Here the harder, more resistant to erosion Kaibab Limestone of Permian age rises to the surface to form the broad plateau that continues westward nearly to Hoover Dam. All younger rocks have been stripped from this geologi-cally high country, except for rocks formed here and there by very recent volcanic eruptions. Major Powell noted this abrupt upfold in the rocks, and also a second fold, the East Kaibab monocline, that the river crosses near the mouth of the Little Colorado River to bring ancient metamorphic rocks to the surface and form Grand Canyon proper (see figure 18). Thus, in two great stair steps, geologic forces have created the uplands through which the Colorado River has been forced to digress.

CHAPTER THIRTEEN

THE EAST KAIBAB MONOCLINE   This flexure of Paleozoic sedimentary rocks is draped across a basement fault, as seen from the air above the eastern Grand Canyon. The mouth of the Little Colorado River is at the lower right of the photograph, and lower Marble Gorge is seen the upper right.

Major Powell believed that the land was being uplifted at the same rate at which the river was down-cutting—the river was able to maintain its drop toward sea level as the rocks were being folded and the canyon was being deepened. However, it is now known that the folds are much older than the river, and Powell could not h ave been correct. The last movement on the folds occurred about 65 million years ago at a time when marine waters covered this region. For the past 30 million years or so, after the rivers had taken stable courses, down-cutting has exceeded uplift.

Some geologists who worry about such things have since believed that the early course of the Colorado River was diverted toward the south when the ever-deepening channel encountered the harder rocks at the East Kaibab fold. They thought that the river followed the general course of the present-day Little Colorado River southward to form a lake, Lake Bidahochi, into which sediments were deposited in the Hopi Buttes country of east-central Arizona. Meanwhile, they postulated that

GEOLOGY IN THE GRAND CANYON   131

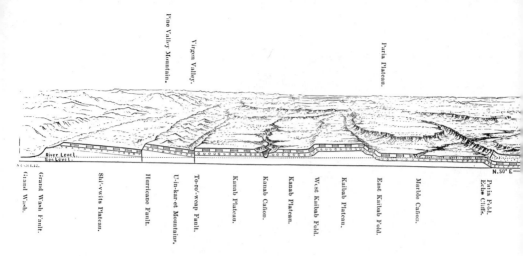

FIGURE 18   Schematic Block Diagram Showing the Main Structural Features Exposed in Grand Canyon. Major John Wesley Powell, who first travelled through the canyon by boat in 1869 and again in 1872, created this interpretive scheme. From Powell (1875)

a separate river system was draining the Kaibab Uplift toward the west, and with the natural process of headward erosion, eventually dissected the highland and captured the Colorado River, diverting it into its present-day westward route to the Gulf of Mexico.

Another possibility has been proposed that the old course of the Colorado River followed its present-day path as far west as Diamond Creek and then turned southward along the course of that tributary. Eventually underground drainage was established through a cave network in the Redwall Limestone, and when the rocks of western Grand Canyon were sufficiently undermined, they collapsed to form the western canyon.

A still more recent interpretation has the original Colorado River flowing westward along its present course to Kanab Creek and then flowing northwestward to some unknown destination in the vicinity of the Uinkaret and Shivwits plateaus. It was later captured by the headward erosion of a west-draining river system to form its present western course. It would seem that wherever river and/or lake deposits of later Tertiary age are found, a course for the ancient Colorado River can be postulated. Conclusive evidence in the Lake Mead area now confirms that the present-day western exit of the Colorado River and its direct connection with the erosion of the Colorado Plateau occurred between 4–6 million years ago, however the river may have got to that location (Ivo Lucchitta 1990).

THE GORGE OF THE LITTLE COLORADO RIVER   This aerial view looks south into the gorge just above its confluence with the Colorado River in the Grand Canyon. Rimrock of the canyon is the Kaibab Limestone of Permian age with sedimentary rocks of Paleozoic age exposed in the canyon depths. The San Francisco Mountains, the sacred mountain of the west, are recent volcanic peaks seen in the far upper left or south.

MARBLE CANYON

The countryside at Lees Ferry is open; the valley is relatively broad. The Colorado River emerges from the Echo Cliffs to the east, where colorful rocks of Mesozoic age form hogbacks that guard against passage in that direction. Older rocks, the gray cliffs of the Kaibab Limestone, begin their rise to the west again almost immediately, to form impassable vertical canyon walls. Here is where an ancient Indian trail forded the river for hundreds of years; here is where Fra. Silvestre Vélez de Escalante and his band of explorers crossed the river in 1776, the natural crossing only to be rediscovered by Jacob Hamblin, the Mormon scout and back-country guide, in about 1864; here is where John D. Lee, hiding from the law for his part in the Mountain Meadows Massacre, built a ferry in 1871 to facilitate communications between Mormon country to the north and Indian territory to the south; here, because of its access, is the starting point of all Grand Canyon river expeditions.

The valley at Lees Ferry is open and the river is accessible because the sharp flex in the Echo Cliffs monocline has inadvertently brought the soft, readily eroded Triassic Chinle and Moenkopi shales to the surface, providing a convenient topographic break between vertical cliffs in Glen Canyon upstream and Marble Canyon below. The dark reddish brown slopes at river level are formed on the Moenkopi Shale, its erosional vulnerability made obvious by the many "hoodoos" seen along the entry road. These formed where boulders fallen from the cliffs of Moenave Sandstone protected pinpoints of shale from weathering where they came to rest, and goblinlike pinnacles have resulted.

A trip down the Colorado River through Marble Canyon and on into Grand Canyon is a trip back through time extraordinary. Because the rock layers dip gently toward the east and the river is flowing downhill toward the west, rocks encountered at river level become older and deeper into the Earth's crust as the trip progresses. Each layer that rises to river level in progressive order is necessarily older than those above until the river crosses into highly metamorphosed crystalline basement rocks in Granite Gorge that date at about 1.8–2 billion years old. The Kaibab Limestone that forms the rimrock of the canyons is the first cliff-forming rock layer seen just below Lees Ferry. It is Middle Permian in age, or about 270 million years old. Thus, one takes a journey back through 1.5 billion years of Earth history, one year at a time (see figure 19).

In less than a mile below the boat ramp at Lees Ferry, the Kaibab Limestone emerges to river level and immediately forms vertical cliffs along the river. The Kaibab Limestone is significant in that it is highly resistant to erosion, and thus forms the rimrock of Grand Canyon for the next 280 miles, blocking views of the younger, more colorful Mesozoic-age rocks for the remainder of the length of the canyon. The formation is thin here, as one is near the eastern limits of deposition of the shallow marine deposits, and in only another 1.5 miles the underly-

CHAPTER THIRTEEN

LOWER MARBLE GORGE   This aerial view looks eastward, just upstream on the Colorado River, from the Grand Canyon. The river has carved this abrupt canyon into sedimentary rocks of Paleozoic age. The Vermillion Cliffs, consisting of sedimentary rocks of Mesozoic age, may be seen in the upper left.

ing Toroweap Formation rises to river level. The two, predominantly limestone formations are difficult to separate here as they appear similar in composition and form nearly continuous cliffs. As they both thicken to the west, they become noticeably dissimilar. The lower Toroweap Formation contains more shale beds and breaks into ledgy slopes beneath the skyline cliffs of the Kaibab. The two formations are separated by an erosional surface, formed between the deposits of two distinctive advances of the sea in Middle Permian time.

Noticeable cross-beds in the Coconino Sandstone are seen at river level as the river passes below Navajo Bridge at Mile 4.5. This formation was deposited in windblown dunes and thickens to form massive, light-colored cliffs in Grand Canyon proper downstream. In this part of Marble Canyon, where the cliffs are nearly vertical and the formations are thin, the Kaibab, Toroweap, and Coconino formations are

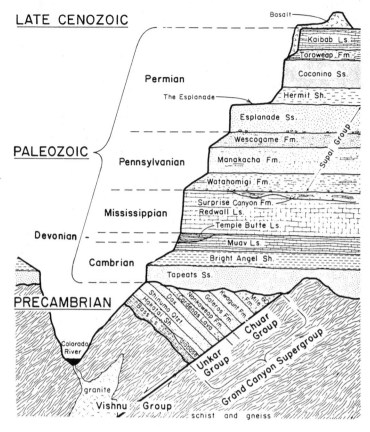

FIGURE 19 A Columnar Section Showing Rocks Exposed in Marble and Grand Canyons. The granitic body shown intruding the Vishnu Group is the Zoroaster Granite. Modified from Potochnik and Reynolds (1990)

difficult to separate visually, but they soon form distinctive separate cliffs on distant skyline views to the west. Then, almost immediately, reddish brown slopes of the Hermit Shale give color and some width to the canyon walls. The bright reddish brown shale and mudstone represent deposits of fine-grained sediments on coastal mudflats and adjacent lowlands.

RAPIDS

The first significant rapid in Marble Canyon is at Badger Creek, so named because Jacob Hamblin once shot a badger here. As with nearly all rapids in Marble and Grand canyons, Badger Creek Rapid formed when heavy rains, perhaps a flash flood, washed debris-laden mudflows

An Aerial View of Glen Canyon Dam and Lower Lake Powell
Page, Arizona is just out of view to the right. Rocks in and near the canyon are
the Glen Canyon Group of Jurassic age.

down the tributary canyon of Badger Creek, depositing the slurry in the
river to form a partial dam. Within a short time the river easily removed
the mud matrix, leaving behind boulders of all sizes that restrict the
flow of water, thus forming rapids. Prior to the construction of Glen
Canyon Dam when annual flood waters sometimes reached flows
greater than 200,000 cfs, boulders the size of trucks were often rolled
around, even removed, making the rapids shift in detail on an annual
basis. In this way, new rapids can form, and former major rapids can
diminish in stature or even disappear.

Only once since Glen Canyon Dam was closed in 1965 has the river
flow exceeded 100,000 cfs. Exceptionally heavy snows in the Colorado
and Wyoming Rocky Mountains melted rapidly in the spring of 1983,
catching Lake Powell above the dam by surprise with a nearly full load
of water. The United States Bureau of Reclamation was trying to fill the

NAVAJO MOUNTAIN This structural dome, formed by the intrusion of igneous rocks at depth, guards Lake Powell in rugged western Navajo country. Rocks widely exposed here are in the Glen Canyon Group of Jurassic age, although Navajo Mountain is capped by the Dakota Sandstone of Cretaceous age.

reservoir to capacity—and did! The massive water buildup could not be discharged fast enough from the flood gates, and water overflowed the top of the dam; the height of the dam was extended by adding walls of plywood sheets at the top. The dam was nearly destroyed as flood-waters reached flow rates of from 90,000–110,000 cfs for several weeks in Marble and Grand canyons below.

Crystal Rapid at Mile 98.1, formed as recently as 1966 by heavy rains on the North Rim and consequent mudflows in the tributary canyons, was studied intently by hydrologists during this flood of 1983. Because of its youth and controlled low flow rates since its formation, the rapid had not been fully stabilized. Boulders that had formed boat-swallowing holes in mid-rapid were moved aside and huge standing waves developed at the head of the rapid. Forty-five-foot-long J-rig river boats were flipped end-over-end at peak times of flow. When the flooding subsided, it became clear that Crystal Rapid had been shifted entirely to the left side of the river, and the right half was an easy, smooth run. Other older, more mature rapids were also noticeably affected, but not to the extent as Crystal.

CROSS BEDS

Just downstream from Soap Creek Rapid (Mile 11.2), light-colored, more resistant rocks of the Esplanade Sandstone rise to river level. Soap Creek, entering from the north, was named when the badger shot by Jacob Hamblin in the canyon just upstream was brought here and cooked overnight; the water he boiled it in was so alkaline that the grease from the badger had turned to soap by morning. The canyon again narrows, with near-vertical walls that display a peculiar variety of swooping, curved bedding. Flat-lying ledges of sandstone separate cliffs made up of bundles, or sets, of bedding that is inclined in varying angles and directions. This is a peculiar type of cross-bedding that is unlike that of the Coconino Sandstone seen under the bridge, causing considerable anguish to those who ponder the origins of layered rocks.

Cross-beds form when sand is transported by a current of either wind or water, and drops out of circulation on the lee, or downstream, protected side of a sand bar or dune. Current ripples form in the same way, only on a Lilliputian scale. The sand grains fall onto this back slope and pile up until the slope reaches the angle of repose (the slope at which angle the grains begin to slump downslope). That angle is steeper in air than in water. Thus, beds of sand are formed at angles of varying steepness, indicating also which direction the current was moving at the time and place the sand was deposited.

Cross-beds in windblown (eolian) deposits tend to be larger in scale, with thicker sets of cross-beds inclined at higher angles than those formed under water. The Coconino Sandstone, at least in Marble and Grand canyons, is a classic example of sand that was deposited in an eolian environment; just about all geologists tend to agree with that interpretation.

Because there are so many cross-bedded sandstones in the world, modern dune sands have been studied intensely throughout the world. There are unlimited types of dunes and limitless varieties of deserts on a global scale. However, underwater sand deposits are never studied in three dimensions internally, because no one has discovered a way to dig vertical-walled ditches in sand deposits under water. Therefore, no sensible geologist ever interprets a cross-bedded sandstone to have been deposited in water, other than stream deposits which are quite different anyway. With this background, it is easy to say that even these very peculiar cross-beds in the Esplanade Sandstone, ever so much different from self-respecting dune sands, were deposited in a windblown environment.

The Esplanade Sandstone, whatever its origin, was named because it holds up broad, flat benches in western Grand Canyon where the Hermit Shale and overlying rocks have been stripped far back from the canyon by erosion. The distinctive formation, assigned officially to the upper part of the Supai Group, was deposited during Early Permian time, or about 280 million years ago.

LOWER SUPAI

Just below Sheer Wall Rapid, at Mile 15.0, rocks of the lower Supai Group appear at river level. These consist of three named formations, indistinguishable without a lot of professional assistance or blind faith: the Wescogame, Manakacha, and Watahomigi formations in descending order. The Havasupai Indian names were derived from Havasu Canyon. The three formations are distinguished on the basis of unconformities (erosional surfaces) that are difficult to find. All consist of thin bedded sandstones, siltstones, shales, limestones and dolomites deposited in shallow marine waters in Pennsylvanian time. For practical purposes, the lower Supai Group constitutes the ledgy, more slope-forming section lying between cliffs of Esplanade Sandstone above and the nearly vertical walls of Redwall Limestone below.

REDWALL

What a marvelous change in the character of the canyon that unfolds slowly after the top of the Redwall Limestone rises to river level at 23 Mile Rapid. The canyon walls close in to form ghostly, unclimbable bastions; the lighting becomes moody as varicolored shades of pinks and dusky grays digress the silent abyss. The inner gorge slowly deepens into a corridor of somber beauty. Near the water's edge, the rock is highly serrated and fluted, buffed to a glassy sheen by eons of the polishing effects of silt-laden waters. There, too, the rock is a sullen gray, changing upward at the high-water mark to varying hues of pink and red—a veneer of rain-washed mud from the red rocks of the Supai and Hermit world above. This is the heart and soul of Marble Canyon.

These traits mark the origins of the terminology used here. Powell was so impressed by the highly polished surfaces on the rock faces that he was reminded of decorative marble, used extensively in government buildings, railroad terminals, and grand hotels of the time; thus, he named it Marble Canyon. G.K. Gilbert named the formation the Redwall Limestone in Powell's time for the red stain on the cliffs of otherwise gray rock that comprises the most noticeable and recognizable formation in Grand Canyon. We could not use such a descriptive term for a formation name today for the rules state that a formation must be given the name of a geographic feature. Gilbert didn't know about that.

Edwin D. "Eddie" McKee, the grand old man of Grand Canyon geology, published tomes on every rock formation in the canyon. When he studied the Redwall Limestone, he recognized and named four members distinguished largely on the basis of the topographic expression of each. The upper member, seen first on a trip down Marble Canyon, was named the Horseshoe Mesa Member; it erodes to form ledgy cliffs at the top of the massive Redwall and is noticeable for its many caves and caverns formed by underground drainage, both in Mississippian and Recent times. It does not require great genius to note that had the dam been constructed at Mile 40 downstream, it would have been difficult to fill considering the number of caverns along the canyon walls. The massive, vertical canyon walls that distinguish the Redwall Limestone were named the Mooney Falls Member, which was deposited on an erosional surface that can be traced across the entire Colorado Plateau, formed when the Mississippian seas withdrew from the region for a short time. Below, the ledgy, cherty beds of the Thunder Springs Member slowly rise to river level, followed by the low cliff-forming dolomite beds of the basal Whitmore Wash Member. The entire Redwall Limestone was deposited in warm, shallow seas during Early Mississippian time.

DEVONIAN

The Redwall Limestone lies directly on limestone and dolomite of Cambrian age where those first emerge at river level. But what became of the Devonian, Silurian, and Ordovician Systems? A great deal of geologic time is missing here—perhaps 150 million years! As far as anyone can tell, rocks of Silurian and Ordovician age are missing from the Colorado Plateau Province and beyond into the Southern Rocky Mountains, having never been deposited or if ever present were removed by erosion from this vast region. However, to the wary eye, channels filled with reddish dolomite of the Late Devonian Temple Butte Limestone can be seen at the top of the Cambrian formations, beginning at Mile 37.6. Numerous ancient erosional channels, no doubt somehow interconnected, are present on this surface throughout the remainder of Marble Canyon, having been caught up between the unconformity at the top of the Cambrian rocks and the erosion surface at the top of the Devonian. In Grand Canyon, beyond the mouth of the Little Colorado River, these discontinuous deposits gradually become continuous beds that thicken slowly westward until they reach a thickness of about 1,200 feet in the western canyon. The Temple Butte Limestone was deposited in shallow marine water in Late Devonian time, some 365 million years ago.

CAMBRIAN

The shoreline of the great sea that lay to the west in what is now Nevada and western Utah, the Cordilleran seaway, began slowly but surely to encroach upon the land sometime around 570 million years ago, the beginning of Cambrian time. At that advancing shoreline, sandy beaches formed that extended from south of Grand Canyon at least as far north as Montana in a continuous belt. As sea level rose relative to land level the beach advanced slowly eastward, leaving behind a legacy of beach sand deposits as it cannibalized sand from the weathering of previously formed rocks that it traversed, then buried. Because the beach took a few million years to dawdle across the stretch of countryside now occupied by Grand Canyon, the deposits are younger to the east and older to the west. We know that preserved, migrant beach as the Tapeats Sandstone.

Seawater gradually deepened as the beach departed eastward. Consequently, wave action waned and finer-grained sediments could be deposited. Mud and silt were being deposited to the west as beach sands were forming to the east. Of course, this process was very gradual and intermittent, so that beach sands grade upward and westward into mud layers, and a great deal of interfingering of sand and mud deposits occurred. The muddy layers gradually compacted with time to form what we now call the Bright Angel Shale. The Bright Angel Shale overlies the Tapeats Sandstone and is thus younger, but the shale was forming westward simultaneously with Tapeats Sandstone deposition to the east. The occurrences of trilobite species—the little crablike fossils—document these changing ages of deposition.

Finally, at least in Cambrian time, seawaters gradually shallowed and stabilized, and lime mud deposits blanketed previously deposited sediments to form the Muav Limestone. Many bedding characteristics in the Muav Limestone indicate that waters quietly shallowed to inter-tidal depths, and broad tidal mud flats prevailed. These mud flats trailed behind the deeper conditions of Bright Angel Shale deposition, again producing gradational and interfingering rocks that become younger to the east. Together, the Tapeats Sandstone, Bright Angel Shale, and Muav Limestone form the Tonto Group. All units thicken and become older to the west.

Another layer of dolomite overlies the Muav Limestone, topping off the Cambrian System in Grand Canyon. Surprisingly, after a century of study and a profusion of naming rock layers, this one remains nameless. Eddie McKee called it an "undifferentiated Cambrian dolomite." Others have called it "unclassified dolomite," while someone else referred to it as the "Supramuav." Like the Muav, this unit appears to be largely tidal-flat carbonate mud deposits and thickens westward. The age relationships are unknown because fossils have not been found for dating purposes, but it is presumed to be Late Cambrian in age.

The rocks of Cambrian age first appear in Marble Canyon at Mile 37, forming cliffs of the Muav and "unclassified dolomite" beneath the Redwall Limestone, then ledgy slopes of the Bright Angel Shale, followed finally near the mouth of the Little Colorado River by the soft, rounded ledges of the distinctive Tapeats Sandstone. Because of irregular cementation in the Tapeats, quaintly eroded niches and hollows often personalize the outcrops.

LIQUID IN THE ROCK

Rocks, especially those beneath the water table, contain water; the water contains varying amounts of dissolved rock. The amount of water contained in a layer of rock is dependent upon the volume of pore space in the rock; the amount and kind of dissolved rock in the water depends largely on the kind of rock in which it resides. Water will run downhill within the rock if it can.

Fractures within the rocks make good spaces for water to fill. Looking at the rock walls in Marble Canyon would make one think that all rocks are fractured, and that's about right. There is more microscopic pore space between grains in a sandstone than one would think, and it is usually intricately interconnected to permit the water to move, if slowly. Limestones are different, usually having little natural pore space. However, if they are attacked by rainwater, which is invariably a little bit acidic, the calcium carbonate of limestone dissolves readily. Underground drainage often results, and caves, such as those found in the Redwall Limestone in Marble Canyon, are the channels for subsurface runoff.

The lime that is dissolved from the limestone emerges from the rock in solution in groundwater via springs and seeps, where evaporation causes it to be precipitated again in the form of travertine deposits. Travertine, a white, brittle coating on the canyon walls and tributary stream beds, is common in tributary canyons that are fed by springs and seeps, especially those discharging from the Redwall Limestone. One of the springs that feeds the annual flow of the Little Colorado River gushes calcite-laden, thus sky blue, water in a nearly closed travertine dome that the Hopi Indians believe to be the *Sipapu*, or opening through which ancestral Hopis emerged from the depths of the Earth. Other especially obvious travertine deposits decorate Havasu Creek, the Pumpkin Spring, and Travertine Falls in Grand Canyon. Lime-laden spring water gurgling from the Toroweap fault just below Lava Falls has produced extensive travertine deposits that encrust the south canyon walls. Deposition of travertine in Havasu Creek is so relentless as to form tiny dams of cemented grass and twigs frozen in place, remaining green with life. Former seepage from the Bright Angel Shale in lower Marble Canyon has cemented the talus slopes along the river into surface rock that is just now being removed by modern erosion, giving the lower slopes a diseased appearance.

Salt, or halite (NaCl), is also common in groundwater, especially that which has been deeply buried below the surface for long periods of time. When saline waters emerge at the surface in springs or seeps, salt also may be deposited as bitter-tasting coatings and stalactites on the rocks. Such salt-encrusted rocks may be seen along the river at the contact between the Tapeats Sandstone and the underlying Precambrian sedimentary rock sequence, such contacts being a ready host to seeping groundwaters a short distance below the mouth of the Little Colorado River. There the Hopi Indians collect salt for religious purposes on annual treks.

THE GREAT UNCONFORMITY

The National Park Service, in all its wisdom, must name everything, and so the eroded surface beneath the Paleozoic sedimentary rock sequence in Grand Canyon has been named The Great Unconformity, although most park personnel don't know what the word "unconformity" means. An unconformity is a surface formed by the erosion and removal of an indeterminate thickness of preexisting rocks, representing a significant and determinate lacuna of time lapse. Such a surface is made obvious by the salt seeps at the base of the Tapeats Sandstone downriver from the mouth of the Little Colorado River. There sedimentary rocks of the Nankoweap Formation of Late Precambrian age were first deposited about a billion years ago then tilted along an ancient fault and eroded to a nearly smooth plain upon which sands of the Tapeats were deposited some 550 million years ago—and this is only the beginning.

As one floats downriver from Mile 63 below the first exposure of The Great Unconformity, the canyon widens, one enters Grand Canyon proper, and the vista becomes a flame of red rocks; all bedding below the Tapeats Sandstone is inclined, or dips, toward the east along Precambrian faults. About 13,000 feet of Precambrian-age sedimentary rocks, mostly red, have been down-faulted and preserved from erosion beneath the Paleozoic cover rock. The sedimentary rocks were deposited horizontally then tilted by faulting and finally planed by erosion to form an "angular unconformity." The river crosses the culprit fault, the Butte fault, near Lava Canyon Rapid, exposing great expanses of the ancient red-bed deposits—the Dox Formation—to view. The upper 6,800 feet of the younger Precambrian Grand Canyon Supergroup, the Chuar Group, cannot be seen from the river, but is exposed in a down-faulted block to the north, accessible via Lava Canyon. The lower 5,800 feet of the Precambrian sedimentary section is crossed rather rapidly by the river between Lava Canyon and the foot of Hance Rapid.

In the sequence encountered along the river, in inverse order of deposition, these are the red beds of the Dox Formation capped by the black Cardenas Lavas, the cliff-forming hard sandstones of the Shinumo Quartzite near Nevills Rapid, and the red beds of the Hakatai Shale, laced with igneous dikes that fed the Cardenas Lavas about a billion years ago, at the head of Hance. Immediately below Hance Rapid is the

Bass Limestone, intruded by a diabase igneous sill that baked the limestone to form an asbestos seam that John T. Hance tried to mine. It overlies a coarse-grained conglomerate, called the Hotauta Conglomerate, which in turn lies directly on basement rock of the Vishnu Schist in Upper Granite Gorge. In all, the upper Precambrian section appears as fresh and well preserved as Mesozoic rocks so common to Navajo Country, even though these rocks are nearly a billion years older. They are only preserved in Grand Canyon where the section has been downfaulted and thus preserved from erosion below The Great Unconformity. As the Grand Canyon Supergroup is dipping eastward and The Great Unconformity forms a relatively flat surface, the entire 13,000 feet of Precambrian sedimentary rock is bevelled to its base high above Sockdolager Rapid, and the Tapeats Sandstone there lies directly on basement rocks. Thus, the unconformities at the base of the Paleozoic section and at the base of the Grand Canyon Supergroup merge to form one erosional surface.

Which of the two unconformities is The Great Unconformity? By definition, it is the erosion surface at the base of the Paleozoic section; some call the lower surface at the top of the basement the "Greatest Unconformity."

At this point on the river one leaves traditional Navajo Country. The canyons downstream are too formidable for any kind of farming or grazing, with near-vertical rock walls that rise to the sky. Although a fine place to hide if Kit Carson comes after you, this is no place to call home.

THE DOMAIN OF VISHNU

It is exactly at the foot of the treacherous Hance Rapid that the river enters Upper Granite Gorge, a dark, somber, nearly vertically walled chasm that reaches 1,000-foot depths beneath The Great Unconformity. The world closes in as the ancient metamorphic rocks rise from the river's edge. The rocks of the canyon walls are not only sinister in appearance but have the look of gnarled and wrinkled old age. Indeed, age dates average 1.8–2 billion years; that was when the geological clock was last reset.

These ancient rocks, generally referred to as the Vishnu Schist, were once deposited as sedimentary layers, interrupted by lava flows some 2 billion years ago. Then through unknown hellish events these layered rocks were nearly remelted and deformed into the contorted mass seen

IMPENDING STORM OVER MONUMENT VALLEY This traditional Navajo homestead consists of *(from left to right)* a mud-covered hogan, summer brush shelter (wickiup), and sweat lodge snuggled beneath a backdrop of The Three Sisters. The cliffs and spires are in the DeChelly Sandstone, rising above slopes of Organ Rock Shale, both Permian in age. (Photograph by Donald Baars.)

Typical Monument Valley Erosional Remnants *From left to right*: Stagecoach, Bear, Rabbit, and The Castle rise majestically above the valley floor. The monoliths consist of the DeChelly Sandstone above a base of brown Organ Rock Shale (Permian), with occasional thin caps of the Triassic Moenkopi Shale. (Aerial photograph by Adriel Heisey).

THE YEI-BE-CHEI DANCERS These deities in Navajo mythology, appear out of the darkness of Monument Valley at sunrise. The dancing Holy People here consist of DeChelly Sandstone spires above basal slopes of Organ Rock Shale (Permian). (Aerial photograph by Adriel Heisey.)

THE TOTEM POLE   This majestic spire rises above southern Monument Valley at sunrise. Its consists of the DeChelly Sandstone atop a base of Organ Rock Shale, both Permian in age. (Aerial photograph by Adriel Heisey.)

WINGATE SANDSTONE BUTTES   Located in Red Rock Valley of northeast-ernmost Arizona, this geological formation hosts a little-known remote natural arch. Shiprock is barely visible in the upper-right distance, as is another dia-treme in the upper center. The Abajo Mountains are being drenched with "male rain" on the far skyline, with the Carrizo Mountains standing in the upper left. (Aerial photograph by Adriel Heisey.)

SPIDER ROCK, HOME OF SPIDER WOMAN   This mythological figure, who taught the Navajos the art of weaving, consists of about 800 feet of DeChelly Sandstone near its type section in Canyon de Chelly National Monument. Navajo families farm the canyon floor. (Aerial photograph by Adriel Heisey.)

SHIPROCK   Located in northwestern New Mexico and the hardened neck of a Tertiary-age volcanic blowout vent (diatreme), Shiprock rises some 1,800 feet above the dark gray Mancos Shale that constitutes the plateau surface. The southwestward-extending igneous dike to the right and above Shiprock is one of three prominent radiating dikes associated with the once-violent eruptive vent. Monster Slayer, born of Changing Woman, killed the much-feared Bird Monsters in Navajo legend here. (Aerial photograph by Adriel Heisey.)

THE GREEN KNOBS   Rising north of Navajo, New Mexico and lying in the foothills of the Chuska Mountains, these knobs consist of Lapilli tuff within a volcanic neck that was intruded into the Triassic Chinle Formation, with cliffs of Cow Springs-Entrada Sandstone to the right. Green sand for Navajo sandpaintings is obtained from the Green Knobs. (Aerial photograph by Adriel Heisey.)

today in the depths of Grand Canyon then shot through with granitic dikes and sills known now as the Zoroaster Granite. More contorting hot forces tied the granitic dikes into bowknots and other strange shapes, leaving myriad, indescribable forms. This place is as close to the center of the Earth as anyone can imagine.

Due to the presence of basement faults and associated monoclinal folds, the river crosses and exposes the metamorphic crystalline basement rocks three times in Grand Canyon: first in Upper Granite Gorge and again in Middle and Lower Granite Gorges downstream.

HOLOCAUST

The Colorado River rounds a gentle bend at Mile 178, and is met by a midstream sentinel of black rock—Vulcans Anvil. This peculiar exposure of unexpected lava is apparently the neck of an ancient volcano, marking the proximity of Lava Falls, the grandfather of Grand Canyon rapids and the first noticeable evidence of fiery and noisy events that occurred since most of the canyon had been carved into place by erosion. This exhumed volcanic neck is followed shortly by views of lava that cascaded down the canyon walls only about a million years ago. "What a conflict of water and fire there must have been here! Just imagine a river of molten rock running down a river of melted snow. What a seething and boiling of waters; what clouds of steam rolled into the heavens!" (Powell 1875).

The first flows, seen from Lava Falls, cascaded down the canyon walls from Vulcans Throne, a beautifully preserved cinder cone formed on the canyon rim of the Esplanade bench 3,000 feet above. Numerous other volcanic vents formed along the rim and within the canyon proper, emerging from feeder dikes visible in the canyon walls. More than 150 lava flows have left remnants of basalt that may be seen downstream for 85 miles. These flows formed at least a dozen natural dams, and the reservoirs of some of these extended upriver to above present-day Glen Canyon Dam into Utah. The detailed history of the formation of this fiery display is complex and fascinating. For example, Prospect Dam, formed by lava flows about 1.5 million years ago, filled the canyon to a height of 2,330 feet with 4 cubic miles of basalt; it took an estimated 23 years to fill the reservoir with water back up into Utah, and about 3,018 years to fill the lake with sediment (Hamblin 1990). Other of the 12 known dams were smaller, filling in days rather than years, with volcanic eruptions lasting until only 140,000 years ago.

Although the name would suggest that the rapid called Lava Falls formed by river flow across remnants of lava, that is not the case. Mud flows and associated rock debris from Prospect Canyon from the south have constricted the river's flow to form the most treacherous and most feared obstacle to river travel in Grand Canyon.

Dozens of little-known collapse features pock the plateaus in and around Grand Canyon. These pipes are often highly mineralized, having produced high-grade uranium ores. The Orphan Mine, located a short distance below the South Rim just west of Grand Canyon Village, first produced copper and then uranium when that mineral's value was realized in the 1940s but was mined out and closed only a few years ago. Other pipes in the area outside of the National Park are still productive as the market for uranium permits.

Thus, the geologic history as displayed in the grandest of canyons commenced and ended with fiery violence over an interval of nearly 2 billion years. Only the numerous incursions of marine waters resulted in the sedimentary deposits that intervene to form the magnificent canyon walls. So rich a legacy!

CHAPTER FOURTEEN

NAVAJO BLACK GOLD

O f all the natural resources to be found in Navajo Country, without
doubt the most valuable is petroleum—oil and natural gas. Petroleum is the result of the natural decay and maturation of organic
substances caused by deep burial and subsequent elevated temperatures
over eons of geologic time. Oil and gas are generally found in sedimentary rocks beneath the Earth's surface, accumulated into concentrations
of sufficient quantities to be of economic significance.

Petroliferous juices form in organic-rich, dark-colored shales that
leak into more porous and permeable beds such as sandstone or porous
limestone. At that point petroleum is free to move about in the permeable rocks. It will seep down-dip (down the slope of the strata) if the
rocks contain little or no water, as is the case in the Mexican Hat syncline, or, because nearly all rocks beneath the surface are saturated with
water and oil rises to the surface of water because of its greater buoyancy, it will migrate up-dip (up the inclination of the bedding) in an
attempt to reach the top of the water-laden strata. If the upward migration of the petroleum is terminated in any way, it will accumulate into
oily concentrations within the rock in what are commonly referred to as
"oil pools." The term is somewhat misleading as the oil and gas occur
in the tiny pores between the sand grains that constitute the sandstone
or in tiny pores within limestones, not in underground lakes such as the
term "pool" implies.

The simplest trapping mechanism for stopping oil migration is an
anticlinal fold where the upward inclination of the strata ends and the
beds are bent downward on the opposite flank of the fold. Faults also
commonly form upward terminations of oil migration, as do up-dip
pinchouts of porous rocks, such as the edge of deposition of a bed of
sandstone. These and many other forms of traps permit the concentration of oil and gas into sufficiently large deposits to form commercially
viable oil and gas fields.

It is an axiom of petroleum geologists, especially those who do the field mapping, that the best oil traps occur in the most desolate regions imaginable. Perhaps that is a major reason for oil companies' interest in Navajo Country, at least in the early days of petroleum exploration. Desolation and nearly impossible travel conditions have been trademarks of Navajo Country from the beginning. Until paved highways first appeared on the scene in the 1960s, it was an adventure to attempt travel into Navajo Country even with four-wheel-drive vehicles. Until that time, most exploration was by necessity done on horseback and horse-drawn wagons in typical Navajo style. Field work was accomplished from travelling tent camps. However, petroleum exploration was conducted in Navajo Country, almost from the earliest years following the development of a market for petroleum in the United States. The first attempts to find and produce petroleum were fraught with unprecedented problems, related not only to methods of exploration and drilling, but also to relationships with the landholders, who in this case were the Navajo people and their guardians the Bureau of Indian Affairs.

Many of those problems peaked when the Navajo Nation brought a lawsuit against the United States Government that came to trial in Albuquerque, New Mexico in 1981. Allegations concerned exploration, production, and leasing practices during the period 1920 to 1946, when lawyers for the tribe found and imagined endless ways in which the government had permitted unlawful and unethical procedures that were allegedly detrimental to the Navajo people—the *Diné*. The following discussion of historical events regarding early petroleum exploration and production on Navajo lands is modified and abbreviated greatly from my written testimony for that trial as an expert witness for the United States Justice Department.

THE OCCURRENCE OF PETROLEUM

Oil and natural gas have been known and described in the literature since about the year 3000 B.C. Until the middle of the 19th century, known occurrences were entirely in the form of seeps, springs and even lakes of oil; gas seeps were more rare. Oil was used for hundreds of years for medicinal and purgative purposes for animals and humans alike. Until 1859 petroleum products were obtained from the seeps or from a few wells drilled for water in various parts of the world. The well generally considered to have been the first drilled for oil in the United States was completed on 28 August 1859 at Oil Creek, near

Titusville, Pennsylvania, by the now-famous Colonel Drake. Until about 1920 oil wells were drilled exclusively by drilling near seeps and by the wildcatters' "trends" and "creekology" methods. Because oil seeps usually occur in creek beds or erosional gullies where petroliferous rocks are first exposed, the original wildcatters took this to mean that oil occurred primarily in creek beds and located their wells by "creekology." "Trendology," on the other hand, was often used to locate well sites on lines between seeps or already productive wells. The early practical oil men completely ignored the "dowsers" and "witchers" and especially shunned geologists and their impractical theories.

The Drake well spurred drilling activity in Pennsylvania for several years, and on 20 May 1860, Prof. Henry D. Rogers noted in a paper presented to the Philosophical Society of Glasgow that the new oil fields of Pennsylvania were located on anticlines (upward folded strata). T. Sterry Hunt of the Geological Survey of Canada gave the first clear statement of the "anticlinal theory" for petroleum occurrences in a public lecture in Montreal that was published in the *Montreal Gazette* in 1861. Twenty-two years later, in 1883, I. C. White was (incorrectly) credited with the founding of the anticlinal theory but at least was the first prospector to put the theory into practice.

Oil and natural gas commonly occur on anticlines because petroleum rises to the top of water-bearing porous rocks, such as sandstones, and continues its upward migration until it is trapped by some kind of termination of the path of migration. An anticline is the simplest form of up-dip trap, but faults or changes in the porosity or permeability of the carrier strata also provide trapping mechanisms.

A more general structural theory evolved around 1900, as geologists began to realize that faults, unconformities, and sand pinchouts were equally important in trapping hydrocarbons. By 1908 a few companies were employing geologists to find oil, and geologists were widely employed for that purpose by 1920. It was not until the middle 1920s that geophysical methods were developed in Europe for determining favorable structures beneath the surface. It was not until the early 1940s that seismograph surveys were conducted in the Four Corners area.

The somewhat modified anticlinal theory was in widespread use in the western United States in the 1920s. "The petroleum geologist in 1921 was primarily and almost solely an anticline hunter, and the Rocky Mountain district held many intermontane basins jewelled with beautiful anticlines. . . . The San Juan, Paradox, and Black Mesa basins . . . contain numerous anticlines which stand out prominently in the bare terrain. . . . Several oil companies mapped various anticlines in more or less detail during the early 1920s" (Owen 1975).

These included the Hogback, Rattlesnake, Table Mesa, Beautiful Mountain, Beclabito, Boundary Butte, Chimney Rock and Mancos Creek anticlines in the immediate Four Corners area. Surface mapping of anticlines remained the primary exploration tool in the region until as late as 1952, as my first assignment as a petroleum geologist for Shell Oil Company was surface mapping in the area between Boundary Butte and the San Juan River in that year. The first intensive use of seismograph surveys to locate subsurface anticlines and faults in the Four Corners area began that year, and these were conducted simultaneously with the surface plane table mapping. The Aneth anticline, where the discovery well for the giant Aneth Oil Field was drilled in 1956, was located and leased on the basis of plane table mapping by Robert Breitenstein for Texaco in about 1950.

PETROLEUM GEOLOGY OF THE NAVAJO COUNTRY

The geology of the Navajo Country was almost totally unknown until Gregory began his reconnaissance studies of the region in 1909. Prior to that time, Newberry, a geologist with the Ives Expedition, had travelled the trail from the Little Colorado River via Oraibi to Fort Defiance in 1857–58, as did Marvine and Howell of the 1872–73 Wheeler Survey. Newberry, later a geologist with the Macomb Expedition, described the outcropping Mesozoic strata along the San Juan River from Canyon Largo to Bluff, Utah in 1859. W. H. Holmes of the Hayden Survey crossed the San Juan River and made a study of the Carrizo Mountains, published in 1875. Schrader in 1906 and Shaler in 1907, published reports on the extensive coal deposits along the Hogback monocline, but their reports shed no light on the petroleum possibilities of the region.

Herbert E. Gregory, a geologist with the USGS, conducted field work throughout the Navajo Reservation during the summers of 1909, 1910, and 1913 that culminated in his classic report on the Navajo Country in 1917. In his introduction, Gregory stated that "Satisfactory maps are lacking, roads are few, and trails poorly marked, water is scanty and generally poor, and food for animals is scarce. . . . Geologic field work in such a country is necessarily reconnaissance; some of it, in fact, is exploratory."

Gregory's work was primarily to study water supplies, but he

mapped much of the region in general fashion and made rather extensive studies of the stratigraphy, economic geology and physiography.
His report included the first geographic and geologic maps of the
Navajo Country, but only the larger and more obvious of the geologic
structures were mapped. Although he mapped the area in the vicinity of
the village of Shiprock, he did not show the Hogback, Rattlesnake,
Table Mesa, Tocito or Beautiful Mountain anticlines, but his map symbols do suggest the presence of the more obvious Beclabito and Boundary Butte structures. Thus, by 1917 the major petroleum-producing
structures of the Four Corners country were apparently unknown. He
did, however, describe the discovery of oil at the San Juan Oil Field
(Mexican Hat), the presence of petroleum in surface sand near Baker's
Trading Post, and the non-commercial Seven Lakes Oil Field.

Gregory shed some light on the nature of the potential petroleum
reservoirs of the region in his descriptions of the stratigraphy. His obser-
vations on the nature of the Dakota Sandstone, the primary shallow
reservoir rock in northwestern New Mexico of later years, were quite
accurate, and the described complexities would plague drillers in the
1920s. Gregory also described petroliferous sandstones in the Pennsylvanian Hermosa Group (his Goodridge Formation) in exposures along
the San Juan River canyons. These strata were to become petroleum
reservoir rocks in future deep tests in the Four Corners area.

As represented by Gregory's report in 1917, the geology of northwestern New Mexico consisted of the large, north-trending "DeChelly
Upwarp," now known as the Defiance Uplift, generally lying along the
Arizona–New Mexico border; a broad structural terrace, now known as
the Four Corners platform, bordering the uplift to the northeast; and
the "Chaco Downwarp," now known as the San Juan Basin, lying east
of the Hogback monocline. The relatively small anticlines of Hogback,
Rattlesnake, Table Mesa, Tocito, Beclabito, and Beautiful Mountain
occur on the Four Corners platform lying between the prominent Defiance monocline along the east flank of the Chuska Mountains and the
Hogback monocline east of Shiprock. It was not until the oil companies, especially Midwest and Continental, became interested in the
potential of the area that the smaller anticlines were realized and
mapped in detail by field geology parties around 1920. These so-called
"anticline hunters" were thorough in their search for oil structures,
mapping elevations on key stratigraphic marker beds by plane table
and alidade, and contouring the resulting structures, pioneering the
latest techniques of the time. The thoroughness of their work may be

measured by the fact that every anticlinal structure expressed at the surface on the Four Corners platform and westward to Boundary Butte was mapped and nearly every one was drilled in the 1920s.

That portion of the Navajo Reservation lying north of Boundary Butte in Utah was apparently overlooked by the oil company field geologists prior to 1946. The Boundary Butte anticline is a very large and prominent structure that was mapped in the early 1920s. North of that structure and north of the Arizona-Utah border, the strata dip gently northward into the Blanding Basin with the deeper structural depression lying approximately along the San Juan River. Several anticlines are present on the structural slope, but their presence is almost imperceptible to the eye, and access to the area was extremely difficult as late as 1952. As a consequence, it was not until the exploration surge in the 1950s and the advent of extensive seismograph surveys that the now-productive structures at Akah, Tohanadla, White Mesa, Aneth and others were discovered and drilled.

PETROLEUM DEVELOPMENT OF THE FOUR CORNERS REGION

A few wells had been drilled in the greater Four Corners region in the search for petroleum prior to 1908, but none was successful. The activity centered around the Durango–Pagosa Springs and Farmington area. E. L. Goodridge noticed oil seeps at what is now known as Mexican Hat, Utah, and at the mouth of Slickhorn Gulch along the San Juan River during a prospecting trip down the river in 1879. He filed placer claims on the land adjacent to the Mexican Hat seeps in 1882 and drilled a discovery well that was reported to be a gusher on 4 March 1908. The Mexican Hat discovery led to the development of the San Juan Oil Field, as "twenty-seven wells, nine of which are productive, had been drilled by the end of 1911." Thus, the first potentially commercial oil production had been discovered in the Four Corners country (Wengerd 1955). As it turned out, oil was never transported out of the immediate Mexican Hat vicinity until 1972, as it was uneconomical to haul the small amount of production by wagon. The Mexican Hat (San Juan) Oil Field occurs in a prominent syncline—not an anticline and thus not complying with the prevalent anticlinal theory of the day. The reason is that oil-bearing strata at Mexican Hat lie above the water table, and gravity drainage feeds oil to the structural depression.

The first oil discovery in the San Juan Basin followed shortly on the heels of the Mexican Hat discovery, but by accident. While drilling for water at Seven Lakes in 1911, Henry F. Brock and Jerry Farris encountered flowing oil from a depth of 300–400 feet. Although some 50 wells were drilled in the ensuing "oil boom," the field was never of commercial significance.

The first commercial gas production in the San Juan Basin was discovered just south of Aztec, New Mexico, on 21 October 1921, by the Aztec Oil Syndicate. The well was drilled with little money, a borrowed drilling rig, and essentially volunteer labor, but a pipeline was laid a distance of 2 miles to town and gas was being sold by December of that year. Ironically, unregulated pressures on the line resulted in the destruction of a number of homes by fires.

Neither the Seven Lakes nor the Aztec discoveries were made on surface anticlines, but the successful application of a modified anticlinal theory elsewhere continued to influence the "anticline hunters" in the Four Corners region. Nowels (1929) reported:

> Several of the major oil companies sent geologists into the district . . . and by the early part of 1922 several of the known geologic structures in the Shiprock District had been located and mapped in detail. The Midwest Refining Company applied to the Navajo Tribal Council and secured a lease on the Hogback structure, and the first well was spudded in on August 5, 1922. A 375-barrel flow of high-gravity oil was obtained in the Dakota sandstone and drilling was stopped at a depth of 796 feet on September 25, 1922.

This was the first truly commercial and significant oil production in the Four Corners region.

Almost frantic activity followed the Hogback discovery, and exploration interest was drawn to the nearby Rattlesnake, Table Mesa, Tocito and Beautiful Mountain anticlines (see figure 20). The Navajo Tribal Council and the United States Indian Department initiated a policy of auctioning oil leases to the highest bidder in public auction. The first such auction was held at Santa Fe, New Mexico in October 1923, when the above named structures were offered for sale and most of the leases were sold. Nowels (1929) stated:

> By February 1924, tests were being drilled on Rattlesnake by the Santa Fe Oil Company; on Table Mesa by A. E. Carlton,

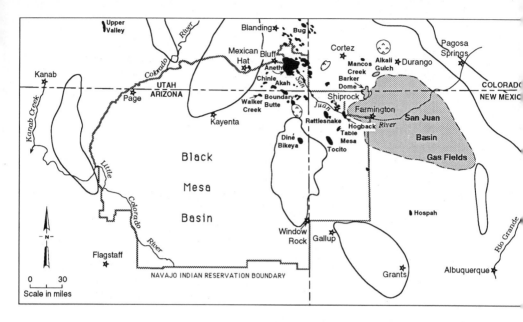

FIGURE 20   Locations of Major Oil and Gas Fields in and Adjacent to Navajo Country. Oil fields are shown in black; gas fields are shown as stippled areas. The most significant of those mentioned in the text are labelled with names here.

with the Producers and Refiners Corporation; on Tocito by the Gypsy Oil Company; and on Beautiful Mountain by the Navajo Oil Company.

OIL DISCOVERIES

HOGBACK OIL FIELD

The first commercial oil in the Four Corners area was at the Hogback anticline, a northerly trending fold immediately west of the Hogback monocline about 10 miles south of the town of Shiprock (see figure 20). The Hogback monocline forms the steep eastern flank of the fold, and the much gentler dipping west flank makes it an asymmetrical anticline. The Cretaceous Dakota Sandstone is the productive horizon at an average depth of about 725 feet. The oil was a very high grade, reported as 63° API gravity.

By 1925, Midwest drilled a total of 12 wells, 7 of which produced commercial quantities of oil; by the end of 1977, more than 5 million barrels of oil had been produced. Drilling of the wells was relatively simple. The completion practices of the 1920s have been described as "drill on in, and stand back!"—a procedure quite primitive by modern standards, but the norm when cable tool drilling was the only method available in the 1920s.

RATTLESNAKE OIL FIELD

The second oil discovery in the Four Corners region came close on the heels of the Hogback Field's success at Rattlesnake Field, some 12 miles west of Hogback and about 7 miles southwest of the village of Shiprock (see figure 20). The discovery well drilled by the Santa Fe Corporation found oil and some gas in the Dakota Sandstone at a depth of 821 feet in early 1924. The company obtained the lease for $1,000 at the first land sale in Santa Fe; Continental Oil Company (Conoco) bought half-interest in the lease for a reported $1,000,000 in the fall of 1924 and became the field operator.

The Rattlesnake structure is an elongate, northwest-trending asymmetrical anticline, with very gentle stratal dips. The structure is the largest of the Four Corners anticlines. The original productive horizon was the Dakota Sandstone at an average depth of 748 feet. A total of 118 wells were drilled on Rattlesnake by the end of 1946; 83 were listed as oil wells. All were drilled by cable tool methods; none of these wells produced after 1966. Two deep tests were drilled into the Mississippian Leadville Formation and produced nonflammable gas containing helium.

The oil discovered at Rattlesnake Field was a very unusual type of petroleum. At 76° API gravity, it was the lightest oil ever discovered at that time. It looked and acted much like kerosene. Indeed, the crude oil was sold and trucked to Cortez, where it was used to fuel public vehicles, such as police cars and fire trucks, without processing. The difficulties in handling such peculiar oil were manifold: the extremely high gravity and high volatility of the oil made it impossible to gauge the tanks by ordinary means and made it almost impossible to move the oil through pipelines by gravity. Further, the lighter hydrocarbon fractions, mainly butane, were lost to boiling and evaporation at surface temperatures. Total cumulative production of Rattlesnake Field as of November 1951 was 4,501,897 barrels of oil.

TABLE MESA FIELD

The Table Mesa Field lies about 8 miles southwest of the Hogback Field on the high crest of the Hogback monocline (see figure 20). It is a northeasterly trending asymmetrical anticline, whose steeper eastern flank is the Hogback monocline. It is a relatively small structure.

The first well drilled on Table Mesa was by Producers and Refiners Corporation and was completed as a water well in 1924. Continental Oil Company became interested in the tract and drilled a well at a new location, and on 1 September 1925 obtained a 325-barrel flow of high-gravity oil from the Dakota at 1,317 feet.

TOCITO DOME

The Tocito structure is a northwest-trending anticline lying about 10 miles southwest of Table Mesa Field (see figure 20). It was considered to be the "most ideal" of the structures in the Four Corners area, and the acreage was acquired by the Gypsy Oil Company at the 1923 Santa Fe auction. The structure was drilled in 1924; both the Dakota and Entrada sandstones contained water. Pan American (AMOCO) drilled the discovery well for the significant oil and gas field producing from the much deeper Pennsylvanian Paradox Formation in 1963.

BEAUTIFUL MOUNTAIN ANTICLINE

Beautiful Mountain is a northwesterly oriented anticline some 8 miles west of Tocito dome. Although it is a well-defined fold, the Dakota Sandstone crops out on the northern extension of the anticline and along an extensive strip some 5 miles to the west on the Defiance monocline. The Navajo Company was formed to drill the structure on a lease obtained at the 1923 Santa Fe auction, finding only a show of oil and fresh water. Helium-bearing gas was discovered at Beautiful Mountain in 1975, when Petroleum Energy drilled to the Mississippian Leadville Formation. The complete lack of oil in the shallow sandstones at Beautiful Mountain and the presence of fresh, soft water indicate that the structure has been flushed by flowing underground water generated locally.

BECLABITO DOME

As translated by Gregory, the word *Beclabito* is Navajo for "spring under a rock." Variously referred to as Biltabito, Bitlabito, Bitabito, Beclabato, as well as Beclabito, the place is about 12 miles northwest of Rattlesnake Field near the Arizona border. The structure is a very sharp dome-shaped fold, with the Wingate Sandstone exposed at its crest. Consequently, the Dakota Sandstone has been eroded from the structure and is not a potential reservoir. The prominent circular shape of the dome plus the red Wingate Sandstone in its core makes Beclabito dome a geographic, as well as structural, landmark in the area just east of the Carrizo Mountains.

Continental Oil Company drilled a well in 1943 that bottomed in rocks of Devonian age, thus testing all of the potential reservoir layers. Oil shows were virtually nil, and the well was abandoned. The dome was again drilled by Pan American Petroleum Corporation in 1967–68, and gas was discovered in the Organ Rock Shale of Permian age; the well was completed as a shut-in gas well.

BOUNDARY BUTTE FIELD

The Boundary Butte Field is a very large northwest-trending anticline that lies about 25 miles west of the Four Corners astride the Utah-Arizona border (see figure 20). It is a very prominent surface structure that has attracted geologic interest since the earliest exploration of the region began in about 1920. The fold is the largest structure east of Comb Ridge in the Four Corners area.

The first well drilled on Boundary Butte was by the Southwest Oil Company located on the apex of the fold. The well was drilled with cable tools, and was spudded in November 1923; drilling ceased in early 1930 in the Pennsylvanian Hermosa Group at a depth of 5,612 feet. Oil was found at a depth of 1,560–1,562 feet in what was believed to be the Shinarump Conglomerate (now known to be the DeChelly Sandstone). It was called "the most important discovery of oil in the Four Corners area." The hole eventually penetrated the Pennsylvanian Paradox Formation, where strong flows of gas (never gauged) were encountered between 4,890–5,608 feet. Because of gas blowouts and mechanical problems in containing the high pressures, the well was abandoned. The second well drilled on the structure was by Continental Oil Company in early 1930; it found oil in the DeChelly Sandstone and gas in the Paradox. This hole was also lost due to mechanical problems and was abandoned.

The official discovery of oil in the DeChelly Sandstone is credited to Western Natural Gas Company in 1948. Since that time, 29 productive wells have been drilled to the DeChelly. By 1977 Boundary Butte had produced more than 4 million barrels of shallow oil. Western Natural Gas Company completed a deep gas well in 1948 that reportedly flowed 25,000 MCFGPD (thousand cubic feet of gas per day) from the Pennsylvanian Paradox Formation; the deep gas has only been used for in-field development drilling.

STONEY BUTTE FIELD

The Stoney Butte anticline trends almost north-south along the eastern border of the 1868 Navajo Reservation. Midwest Refining Company geologists mapped the structure in 1927 and drilled a test well to a depth of 3,063 feet in 1928, finding small amounts of oil in the Menefee Formation; the well never tested the Dakota Sandstone.

The discovery well was drilled by Southern Union Gas Company in 1950. It produced some oil from the Dakota Sandstone then was deepened to basement and abandoned in 1952. No oil or gas shows were encountered below the Dakota Sandstone. Oil was found about a mile to the north along the anticline in the shallower Menefee Formation in 1953. Three wells had produced a total of 15,851 barrels of oil when the field was abandoned in 1957.

DISCOVERIES OFF NAVAJO RESERVATION, PRE-1946

Oil and gas discoveries made between 1920 and 1946 not on the Navajo Reservation but within the Four Corners area are few and for the most part relatively insignificant. Of these, only two fields, the Mexican Hat (San Juan) and the Mancos River, are close to the Navajo Reservation boundaries, and both of these fields are truly insignificant from a commercial point of view (see figure 20). Two of the most important fields discovered during this time interval, the Barker Creek and Ute Dome fields, lie within the New Mexico portion of the Ute Mountain Indian Reservation. The Hospah Field lies within the 1907 boundary of the Navajo Reservation in the area known as the "checkerboard" lands.

POST-1946 EXPLORATION

Following the initial discoveries of the 1920s, mostly in the immediate Four Corners area, exploration activity slowed appreciably. The market was limited for new reserves; prices for petroleum products remained low. Accessibility was a major factor as more readily available petroleum was known to occur in the Midwest and the Gulf Coast states. Pipelines for oil and gas distribution were not practical or warranted in Navajo Country—the market was not ready and would not be enhanced through the World War II era. Anyway, it was no easier to explore Navajo Country than it had been during the 1920s. Roads were of the two-rut wagon trail variety and few in number, often laden with blow sand and seasonally made impassable by mud; most access was by trail only.

The physical conditions in that area explain in part why oil companies were reluctant to explore the region in earlier years. Also, prior to 1952 virtually all drilling was limited to structures mappable at the surface, and they are small and obscure in southeastern Utah. By 1952 it was realized that the margins of the Pennsylvanian Paradox salt basin, which pass through the Utah portion of the Reservation at depth, were potentially conducive to the deposition of fossil "reefs" that were recognized as potential stratigraphic traps for petroleum. This added incentive for exploration was adequate justification to expend the necessary funds and effort to explore that remote area. The idea was to be proven generally correct and led to the development of vast oil production in that part of the Reservation in the following 20 years.

ANETH FIELD

A flurry of mapping by field parties from several major oil companies enlivened northern Navajo Country—to the chagrin of many local *Diné*, during the years 1952–54. Then the drilling began. It seemed that by drilling the mappable surface anticlinal structures, as had become the rule in the petroleum industry, dry holes soon ringed what would become productive fields. Just as the magic was wearing thin, and drilling activity slowed, quite by accident, Shell Oil Company discovered the Desert Creek Field on its second well drilled into the faulted anticline. The structure had been mapped at the surface during 1952 and further delimited by seismic studies the same year. It was not a big discovery, amounting to only about 250 barrels of oil per day, but it was enough to awaken the enthusiasm of the industry. More dry holes ensued.

Even before Shell geologists had the divine premonition that there was oil in southeastern Utah, Bob Breitenstein was mapping the "boondocks" around Aneth Trading Post north of the San Juan River on the Navajo Reservation for Texaco. He documented the presence of an obscure, very gentle anticline, barely discernable at the surface, that prompted Texaco to obtain the necessary lease to drill. Terms of the lease dictated that a well must be in the drilling process to test the petroleum potential of Devonian strata or that production must be established by 1 January 1956, or the lease would expire. In a last-minute scramble, the Texaco No. 1 Navajo C well was spudded late in

1955, with no inkling of what lay ahead. In early February, much to the dismay of the well site geologist, the well drilled into the same layer that was productive in the Desert Creek Field to the south, the Middle Pennsylvanian Desert Creek pay zone, and oil flowed, and flowed, and flowed. Officially the initial production is listed at 1,704 barrels of oil per day, but testing revealed a considerably greater potential. Gradually, new discoveries that followed at McElmo Creek, Ratherford, and White Mesa, all found within a year, merged with the Aneth discovery as development drilling proceeded, and one huge oil reservoir nearly 12 miles in diameter and up to 200 feet thick was realized (see figure 20). More than 1,000 wells were drilled in the Greater Aneth Field of which at least 900 were productive of oil or oil and gas. Drilling at first was limited by the state of Utah to 80-acre spacing (8 wells per square mile); the spacing then was increased in the 1970s to 40-acre spacing.

Reservoir rock was of a new variety, not seen before in this part of the world, of highly porous and permeable limestone consisting of fragmented remains of fossil calcareous algae (later identified as the genus *Ivanovia*), stacked on end like corn flakes. Estimates of potential oil production in Greater Aneth Field range upward to 500 million barrels of oil—truly one of America's giant oil fields. The oil boom of the century was ushered in for Navajo Country. Things would never be quite the same again.

Post-Aneth

Aneth meant oil—Big oil! After all of the skepticism, it was now certain that the Paradox Basin and Navajo Country were words synonymous with oil, and the drilling frenzy began. Now it was not just drilling up the Greater Aneth Field, everything must now be drilled, and quickly, before the buzz wore off. Landmen, those who do nothing but buy up mineral leases—mostly lawyers and would-be lawyers—would spring into action. Somehow Shell Oil Company landmen had accidently acquired all of the state "school sections." Sections 4, 16, and 31 of each township had been set aside by the state of Utah to support public schools, and consequently Shell held potentially productive acreage from the start. Soon all unleased federal acreage within 100 miles of Aneth was under lease to one or another major oil company. Texaco was, of course, in the thick of action, as was Shell; Conoco, Superior, Pan American (AMOCO), Gulf, Stanolind, Davis, Sun, Cities Service, Phillips and others all vied for land ownership.

Then the Navajo Nation made its move. Mineral rights to Navajo lands would be put up for sale at sealed-bid auctions. Rental fees were automatically set at $1.25 per acre at first, so sales went to the highest bidder on bonuses, subject of course to USGS and BIA approvals. Navajo tracts were fixed at 4-section plots; bids were set in highest levels of secrecy; everyone made bids; all tracts in any given land sale were taken, whether or not prospective structures existed. Tracts were sold for bonus bids as high as $3,200 per acre, and generally bids lower than $100 per acre were rejected, regardless of the location. As the buying frenzy hit its peak in later Navajo land sales, the standard 16 percent royalty for oil production was nullified, and royalties as well as bonuses were subject to sealed-bid consideration.

Meanwhile, exploration activities continued in relatively orderly fashion. The Aneth discovery led to the search for other Aneths, but most of this hopeful extension of a potentially productive trend, or fairway, affected southeastern Utah more than it did Navajo Country. Another Aneth never materialized, but several smaller, often lucrative fields such as Bluff, Ismay, Cowboy, and Turner Bluff resulted. Most production was from the next higher pay zone, the Lower Ismay, rather than from the Desert Creek zone of Aneth. Acreage south of the San Juan River to about the Arizona state line was actively explored on Navajo leases. Following the Desert Creek discovery in 1954, Akah Field, originally named North Boundary Butte, was drilled and discovered by Shell Oil Company in 1955. Tohonadla, Gothic Mesa, and Chinle Wash, followed in 1957. Anido Creek was drilled in 1961, and so on. Although production is generally attributed to lower pay zones south of the San Juan, most production has actually come from algal bank deposits in the younger Lower Ismay zone. After an exploration hiatus of some two decades, Chuska Energy Company is again exploring this trend along the southwestern shelf of the Pennsylvanian Paradox Basin in the early 1990s.

DINEH-BI-KEYAH

A most unusual occurrence of oil was discovered in the northern Chuska Mountains by the Kerr-McGee Corporation in 1967. While drilling in sedimentary rocks of Pennsylvanian age, presumably exploring for limestone oil reservoirs similar to those in the Four Corners region to the north, the Kerr-McGee No. 1 Navajo well encountered an unexpected igneous sill that had been intruded into the Pennsylvanian sedimentary section. Technically, the reservoir rock is a syenite, but unlike most exposed intrusive igneous bodies found in the area, it is porous and contains oil. Cores taken from subsequent wells show that the syenite sill contains large gas-bubble holes, much like gas bubbles in vesicular basalt lava flows found elsewhere, as well as intercrystalline and fracture porosity. The oil field lies on the northwesterly plunging nose of the Toadlena anticline that obliquely crosses the northern Chuska Mountains in Apache County, Arizona. The field was appropriately named Dineh-bi-Keyah (meaning Navajo Country), or Navajo Field (see figure 20). Thickness of the sill varies up to 160 feet, with as much as 80 feet containing a high-grade oil at 43° API gravity. The discovery well initially pumped 648 barrels of oil daily, although the 17 producing wells were making a total of only 1,000 barrels per day in 1978.

As oil is almost always produced from sedimentary rocks, having been derived from sedimentary organic material, this maverick field has enjoyed international acclaim as a peculiar possible example of oil that may have originated in the depths of the Earth and emplaced with the intrusive igneous rock. The oil is generally believed, however, to have originated in the sedimentary rocks in which the sill was intruded and subsequently leaked into the porous igneous rock.

The financial bonanza from oil production realized by the Navajo Nation was a godsend to the struggling tribe. Consideration was given to dividing the money on a per capita basis and spreading it among the people equally, as was the tradition among other tribes, mainly in Oklahoma and Texas. Fortunately, that idea was rejected by the Tribal Council, and funds were spent for tribal projects that would benefit all. Schools, hospitals, Chapter Houses, roads, tribal businesses and tribal parks were established in the late 1950s and early 1960s. Individuals did not benefit financially, but the tribe made great advances in providing educational, social and health benefits, as well as business income for the Navajo Nation as a whole. When individual payments finally began in the mid-1960s, the traditional Navajo mode of transportation, the horse-drawn Studebaker wagon, disappeared as if by magic, and pickup trucks swarmed on the Reservation. Truck dealers in nearby towns became financial tycoons. Oil had made its mark in Navajo Country.

NATURAL GAS

A closely related product that may be produced along with oil or may be found in separate accumulations is natural gas. Consisting mainly of methane (CH_3), natural gas is relatively plentiful. In the 1920s, little gas was produced with Four Corners oil, and that which was produced was too volatile or too deep to market. Indeed, a viable commercial market for the evasive gas has always been a very real problem, especially in Navajo Country. Although plentiful, there is no market for it. It cannot be bottled, trucked, shipped by rail, or otherwise distributed except by pipeline, and pipelines are expensive to build and operate. Since the early beginnings at Aztec, New Mexico, there has been an abundance of natural gas to supply nearby towns; Farmington, Durango, Cortez, and even Albuquerque have received ample supplies of gas from San Juan Basin fields as pipeline construction is feasible for these towns from nearby productive areas. Nonetheless, the San Juan Basin is a veritable sea of natural gas with nowhere to go. One of the largest natural gas fields in the United States lies nearly dormant for lack of a market.

As luck would have it, the gas fields lay largely beyond the Navajo Reservation to the east, in the bottom and north flank of the structural depression of the San Juan Basin (see figure 20). If Old NavajoLand, that region east of Farmington, through Aztec and beyond to about Chama, New Mexico—had been included in the Navajo Indian Reservation, the Navajo Nation would again have become enriched, this time with gas royalties. Much of the production occurs from the "Checkerboard" lands east of the Reservation proper, but most is from lands owned by private ranchers, and Jicarilla Apache lands. The so-called "Checkerboard" lands were opened to homesteaders after the boundary to the Navajo Reservation had been established. Navajo people qualified for some of the lands that rarely exceed a section (1 square mile) in extent, but much of the acreage went to others. The lands not homesteaded by whites belong to the Southern Ute Indian Tribe. Considerable tracts were given to the Santa Fe Railroad, which like other railroad companies of the 1800s, was given alternate sections within 40 miles on either side of the proposed trackway to encourage railroad construction. Because of this multiple ownership of the San Juan Basin Gas Field, no riches were realized by any particular group.

Natural gas is a close cousin to oil in that it is produced by the decay of dead and buried organic sources, such as vegetation or microscopic animal life. It often occurs in association with oil, but some is derived as a by-product of coal and may occur separately. Gas, like oil, must accumulate in reservoirs that consist of porous rock. Sandstone makes good reservoirs, and sandstones of Cretaceous age are plentiful in the San Juan Basin. Most gas here is produced from the Mesaverde Group sandstones, the Point Lookout, Cliff House, and Pictured Cliffs sandstones are the primary producers, but the Farmington Sandstone of Tertiary age is sometimes prolific. All of these deposits are widespread throughout the structural basin and crop out in splendid exposures around the basin, especially near Mesa Verde where the group name originated.

If natural gas reacts similarly to oil, and it rises by buoyancy in water-wet sandstone, why is it trapped in the structural depression of the San Juan Basin? Why has it not leaked out at the up-dip outcrops? That is a question that has haunted petroleum geologists for decades. In fact, gas seeps are common in Mesaverde exposures, at least along the northern margin of the basin, and farmers often fire their private gas seeps in winter to warm their cattle.

One possible explanation is that much of the gas is contained by hydrodynamic trapping. Water seeping into sandstone outcrops along the high northern boundary of the basin tries desperately to flow down the inclined porous beds of sandstone into the basin, and in doing so slows the up-dip migration of gas, forcing it into basinal reservoirs.

Another possibility is that stratigraphic traps exist, virtually unnoticed, where trends of porous and permeable sandstone vary laterally with silty or shaly beds that hinder up-dip migration by reducing lateral permeability. That this may be a plausible solution to the age-old problem is illustrated by the empirical fact that the best gas-producing trends occur in northwesterly oriented belts that parallel depositional strike. Note the trends in drilling density on any oil and gas map of the San Juan Basin. Depositional variations, in turn, seem to have been controlled by slight variations in paleobathymetry, slight rises or shelving of the sea floor, caused by rejuvenation of basement fractures. The situation is further suggested by the northwesterly linear trends of oil production from deeper horizons, such as the Gallup Sandstone production in the Bisti Field, and oil and gas from the Deep Dakota Field. This cause-and-effect relationship was drawn to the attention, over boos and hisses, of local petroleum geologists in the 1970s, only to resurface in 1992 as a newly discovered phenomenon.

Whatever the trapping mechanism, the San Juan Basin is a world-class repository of natural gas that has been recognized since the 1920s. The lack of a viable market kept the fact quiet until the Pacific Northwest Pipeline was built in the 1950s, when markets as far distant as Portland, Oregon were serviced, and drilling activity increased markedly. Then, a gas line was built from western Canada to the lucrative Pacific Northwest, and cheaper gas flooded that market. Canadian gas backed down the San Juan Basin pipeline as far as Salt Lake City, and the gas industry again went dormant. Since that time, pipelines began to serve markets in California and the Midwest, and some steady but not prolific production was realized.

Drilling activities in the San Juan Basin have fluctuated wildly depending on momentary market conditions. At first, infilling of the giant field at 320-acre spacing (2 wells per square mile) kept drillers busy. Then came the pipeline boom, associated with an increase of spacing to 160-acre (4 wells per square mile), and the drilling centers of Farmington and Aztec again boomed. In the 1980s, with drilling for coal bed methane, that gas directly associated with coal deposits, and the advent of horizontal drilling, another drilling bonanza occurred in an otherwise dormant economic era for the petroleum industry. Still the nasty reality exists that without a ready, significant market, full potential gas production in the San Juan Basin will not be realized.

NAVAJO COAL

B lack rock. It is not pretty, but it burns. In the process it makes energy and is thus valuable. It can be burned for household heating purposes, but it is far more valuable as a means to generate electricity. It is coal; and there is lots of it in Navajo Country.

Coal is the product of decaying plants, mostly formed in swamps. When plants die, they begin to emit gasses and other by-products found in living matter. Much of the lost material is in the form of hydrocarbons, largely methane, and thus decaying plants produce natural gas and other organic compounds that can eventually become petroleum products. Indeed, much of our petroleum is believed to form from the destruction of land plants. What remains in the end, if all goes well, is carbon; the ratio of carbon to other organic compounds determines the economic value of the resulting coal. The higher the carbon content the better as far as coal is concerned.

Swampy environments are ideal for coal formation as the water is stagnant and there is little sediment available to contaminate the organic residue. There is a paucity of oxygen in such waters, at least in the bottom accumulations, so the organic material is not destroyed through oxidation or by bottom-feeding organisms. Grasses and other swamp plants simply settle to the bottom, begin to smell bad, and with any luck become coal. The first substance formed as the process develops is peat, a generally black, rather loose mat of intertwined plant debris with much of the original moisture removed by natural compaction. Peat will burn but not efficiently, and considerable smoke and ash results. After burial of the peat to sufficient depths by sedimentary overburden, more of the lighter organic compounds are expelled with time and further compaction, and low-grade coals begin to form. Deeper burial and more time form higher grades of coal because more of the lighter organic fractions are expelled by ever-increasing compaction. The higher-grade coal burns more efficiently, giving off fewer unwanted by-products. Coal that is high in sulfur content is not

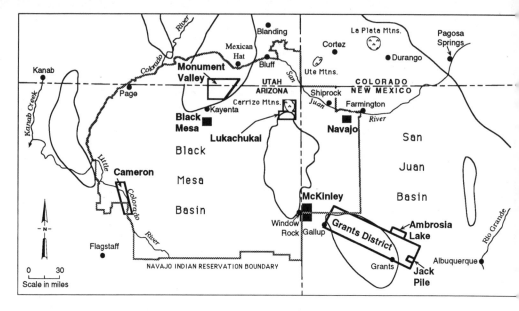

FIGURE 21 Coal and Uranium in Navajo country. Major coal-mining districts are shown in black; open areas within heavy outlines are uranium mining districts in and near Navajo Country.

particularly desirable as most people don't like the rotten-egg-smelling gases that result when the coal is burned; it is also dangerous to one's health. Coals mined in the Rocky Mountain West are generally lower in sulfur than their eastern counterparts, and thus are more desirable for energy-generating purposes.

Most of the coal produced in Navajo Country is from rocks of Late Cretaceous age, as swampy lowland conditions prevailed during times of lowered sea level. As a result, Navajo coal is found almost entirely within the structural basins where Cretaceous rocks have been spared from erosional removal. Also, coal found close to the Earth's surface is the most desirable as it is expensive to remove overburden to reach the coal and even more expensive to dig deep underground mines. The economic value of the coal will not permit such expensive mining operations. Consequently, most coal is mined from the margins of the San Juan Basin where it is at shallow depths, or from the top of Black Mesa where erosion has effectively stripped the overburden from the coal seams by natural processes (see figure 21).

By far the greatest demand for coal comes from the needs of our society for electricity, and these demands are met through the use of coal-fired generating plants. It is far more costly to transport coal over great distances than it is to transmit electrical energy. Consequently,

CHAPTER FIFTEEN

electrical generating plants are necessarily located nearby the source of abundant, inexpensively mined coal. This situation, along with the need for considerable amounts of water, explains the location of the larger coal-fired electrical generators in Navajo Country near Shiprock at the northwestern corner of the San Juan Basin, and at Page, Arizona near large, shallow coal reserves on Black Mesa. Unfortunately, a lack of effective smoke and ash restrictions on these generators at the time of their construction has done immeasurable damage to the visual air quality of the region.

To make matters worse, the only cost-effective means of obtaining Navajo coal has been strip-mining. The process involves wholesale removal of up to 200 feet of soil and rock cover to expose the coal seams, and the removal of tens of feet of coal. Finally, the removed overburden and coal is replaced with fill from ongoing excavations nearby, the surface configuration is restored to a natural appearance, and that land surface is reseeded with ecologically compatible vegetation. Other than the fact that coal has been removed and it is non-renewable, the process seems, at first glance, to be satisfactory. From the Navajo point of view, however, the results are not the same. It is not the land their ancestors grazed and farmed; it is not the land that generations of *Diné* nurtured and loved; it is not the land that contains the bones and ghosts of those gone before. This restored land is different, unnatural, and morally unpalatable. Who could ever call such rebuilt and reshaped land home? What would the ghosts of their ancestors think? What would the Holy People think?

Man's use of coal goes much farther back in history than most of us realize. Prehistoric Indians mined and used coal in their kivas, stone stoves and pottery firing pits on Black Mesa at least as far back as A.D. 1300, perhaps as early as the year a.d. 900, according to Brew and Hack (1939, 8–14) They estimate that production may have totalled 100,000 tons during the period a.d. 1300–1600. Coal was mined and used for local purposes during the 1930s and early 1940s, totalling fewer than 100,000 tons according to records; since 1943, about 10,000 tons per year were taken from 10 underground mines for use in schools and agency facilities on the Reservation, and some was shipped to nearby towns in the vicinity of Navajo Country (Peirce, Keith and Wilt 1970, 15). Since 1926 coal production in Arizona had been less than 300,000 tons until the advent of coal-fired electrical generating plants.

It was not until 1943 that the Tribal Council recognized the economic potential of their vast coal reserves. Some production was realized in the 1950s, and between 1950 and 1958 dollar values received by the Tribe were less than $3,500 per year. By 1960 that amount had increased to only about $13,000 per year (Young 1961).

Then in the early 1960s, Arizona Public Service Company and affiliate interests began construction of the Four Corners Power Plant near Fruitland, New Mexico, between Shiprock and Farmington. Utah International, Inc. opened the coal mine in 1963, with reserves estimated to be more than a billion tons; they leased 31,400 acres of Navajo land to provide the necessary coal for the power plant, estimated to be about 7 million tons per year. Of the 500 employees at the mine, 72 percent are Navajos. The coal is of the sub-bituminous grade that occurs in the Late Cretaceous Fruitland Formation. The mine and power plant, the world's largest such operation in 1977, serve the electrical needs of El Paso, Tucson, Phoenix, Albuquerque and Los Angeles (Karna 1977, 251). San Juan River water is used, based on Navajo rights.

Peabody Coal Company announced in 1966 that it would begin stripping coal from the Wepo Formation underlying the northern parts of Black Mesa. The company announced reserves of at least 350 million tons of strippable coal under lease. Coal was to be slurried through an 18-inch diameter pipeline from Black Mesa to the Mojave Power Plant in Clark County, Nevada, a distance of 273 miles. About 2,000 gallons per minute of water was required for the purpose, derived from 4 water wells drilled into the Navajo Sandstone aquifer. Coal needed to fire the Navajo generating plant at Page, Arizona is hauled by an electric train built for the purpose from Black Mesa.

Economic optimism prevailed in the 1960s when these power plants were built, with little regard for possible environmental consequences. The Navajo Nation would receive more than $2,000,000 a year for 35 years from the Black Mesa strip mines alone, according to Peter Iverson (1981), even at negotiated royalties of less than $0.25 per ton, and much-needed jobs were created. Anyway, nuclear power would soon make coal mining obsolete, or so it was believed, and Navajo coal would then become a useless commodity. Unforeseen environmental deterioration and disappointment in the promised number of jobs, however, soon brought massive protests from the local Diné. Such displeasure over Black Mesa coal-mining operations led to rejection of proposed coal-gasification projects for the Burnham area in the 1970s although royalties offered for the coal had risen to $0.55 per ton (Iverson 1981).

According to Brad Nesemeier, geologist for the Minerals Department of the Navajo Nation in Window Rock, since coal lease renegotiations were completed in the 1980s, coal production has been the principal source of mineral royalty revenue and economic benefit to the Navajo Nation. By the end of 1992, cumulative Navajo coal production had reached 450 million tons. In 1992, for the first time, annual coal production exceeded 25 million tons. Although abundant proven reserves remain on existing mine leases, additional unleased coal resource development is dependent on market conditions, electricity demand, environmental considerations, and the lack of infrastructure (principally railroads) in the western San Juan Basin and around Black Mesa. Scrubbers that will reduce sulfur dioxide emissions by 90 percent will be fully operational at the Navajo Generating Station at Page, Arizona by the year 2000.

CHAPTER SIXTEEN

YELLOW CAKE

The greatest and most efficient source of energy available in the 20th century is nuclear power. It is unfortunate that shoddy engineering and operating practices have saddled the nuclear industry with such disasters as Chernobyl in the Ukraine, and potential disasters such as Three Mile Island in this country. The political fears generated by such events have set nuclear power generation back decades. Then too, the high-level nuclear waste disposal fiasco of recent years has helped eliminate any possibility of developing a viable nuclear power industry in the foreseeable future. Very small amounts of radioactive source, such as uranium, produce enormous energy resources, and uranium is one thing that is abundantly available, especially in Navajo Country.

Uranium was discovered more than 200 years ago, its radioactive characteristics were known by 1896. In 1898, it was realized that radium, extracted from uranium, was useful in the cure of cancer, and a few other uses were soon discovered. Hahn and Strassman successfully split an atom of uranium in 1938, leading to the development of nuclear weaponry by the Manhattan Project that was used by the United States to end World War II. It was not until military use of nuclear energy was eased following the war that control of nuclear reactions was developed for domestic energy needs, and the demand for large amounts of uranium was established. "From 1946 to 1954, the number of uranium mines in western United States increased from about fifteen to over 900" (Peirce, Keith, and Wilt 1970, 104).

Radioactive substances, such as uranium, thorium, radium, potassium, strontium and rubidium, produce energy as particles escape from the nucleus of the atom, forming other elements in the process and releasing heat. In the case of uranium, the most widely used element in energy production, particles consisting of helium are emitted from the nucleus of the "parent" atom to form a "daughter" element that is a form of lead. The rate at which radioactivity proceeds is statistically fairly constant in nature. Consequently, rocks containing radioactive elements can be dated. Age determination involves careful measurements of the amount of remaining parent element in a rock and the resultant daughter element, then multiplying by the "constant" rate at which the process is believed to proceed. What is important in the generation of electrical energy is the heat that is produced in the process.

Nuclear energy for the production of electricity is derived by artificial bombardment of fissionable uranium nuclei by free neutrons, splitting the uranium nuclei violently into two different elements, and producing large amounts of heat in the process. Neutrons released by nuclear fission can bombard other uranium nuclei to set up a continuing chain reaction. If not controlled, enormous explosions can occur, such as those of the atomic bombs, but chain reactions can be safely controlled by the use of neutron-moderating materials. "Theoretically one gram of fissionable uranium could furnish as much heat as three tons of good coal and one pound could supply ten million kilowatt-hours of electrical energy" (Peirce, Keith, and Wilt 1970, 103).

Heat production used for electrical generators from nuclear reactors is far more efficient than heat produced from natural gas or coal, and the process is environmentally much cleaner. The real problem lies in the safe disposal of the spent waste that results. In the fission process, only about 0.1 percent of the original mass is spent in the production of energy. The problem of nuclear waste disposal is far from insurmountable, but political indecision has created a staggering dilemma. Processes and problems, such as those confronting the safety of the now proposed Yucca Mountain high-level waste disposal site in Nevada, are beyond the scope of this discussion and do not directly affect Navajo Country.

Uranium occurs in many different forms in nature. It is sometimes found as a black, waxy mineral called pitchblende that forms as primary intrusive veins, dikes and sills, associated with igneous activity. The mineral is rare in the Rocky Mountain West, where most uranium deposits occur as secondary oxides of uranium found in sedimentary rocks, having been altered from the primary mineral by natural chemical processes that occur at or near the Earth's surface. Secondary minerals of uranium are extremely varied, depending upon myriad natural chemical processes that may occur, and often several minerals are found in combination in any one ore body. The most common, or certainly the best known, of the uranium-bearing minerals is carnotite, a canary yellow powdery mineral that commonly occurs in sedimentary rocks. Many secondary uranium minerals are also yellow, as is the concentrate U_3O_8 known as yellow cake, and the color yellow in rocks is commonly associated with the presence of uranium minerals.

Uranium is ultimately derived from igneous sources but is commonly found in sedimentary rocks following natural oxidation and redistribution by groundwaters. Secondary uranium is soluble in acidic groundwaters and travels through any porous and permeable rock along with the water host. If reducing conditions are encountered by the uranium-bearing groundwater, uranium minerals are precipitated. Reducing conditions, where the oxygen content of the environment is low or lacking, most commonly occur in the presence of organic material, such as fossil leaves or twigs of plants in sedimentary rocks. Precipitation of the uranium oxide may occur immediately adjacent to the organic material or may replace leaves, twigs, even logs, leaving perfect plant fossils consisting entirely of a uranium mineral such as carnotite.

The most common host rock for uranium ores is stream-deposited sandstone, particularly that found in rocks of Triassic and Jurassic age in Navajo Country. Although uranium deposits have been found in other rocks, most production has been from the basal fluvial sandstones of the Chinle Formation, the Shinarump Member in Navajo Country or the Moss Back Member farther north, and sandstones of the Morrison Formation. Because of this, exposures of these formations have been heavily prospected across the Colorado Plateau, having been scarred by bulldozers and shallow shafts to make them readily recognizable from a distance.

Stream deposits of both the Chinle and Morrison formations occur as channels formed by streams that wandered aimlessly across the ancient lowlands. Meandering streams form sinuous channels that are asymmetrical in cross section, cutting sharp banks where the current strikes the outer margin of the curves with erosive strength and slopes where sediments settle to the bottom on the inner part of the curves. Such arcuate sediment accumulations are called point bars for their curved shapes and contain the finer sediment grain sizes along with accumulated deposits of plant debris, like modern-day drift wood. The flow patterns of later groundwater, often laden with uranium minerals in solution, follow the porous and sinuous underground sand deposits until reaching a point bar containing fossil plant debris, where the uranium minerals are precipitated. If not recognized in cross section from natural exposures, point bars may be located in the field by searching for prospecting pits along the outcrops. Commonly, exposures within the prospect pits reveal yellow fossil plant debris replaced by uranium minerals, or yellow halos around fossil plant fragments, verifying the direct relationship between fossil plants and ore localization.

As previously mentioned, the time of crystallization of uranium minerals can be determined by radiometric dating techniques—determining the ratio of parent uranium to daughter lead concentrations. Studies of this sort usually reveal that uraniferous mineralization took place during about Middle Tertiary time on the Colorado Plateau, long after the sedimentary rocks of Triassic and Jurassic age were deposited. Thus, mineralization did not occur during deposition but at some later time. One of several differing interpretations suggests that perhaps mineralization occurred during intrusion and upwelling of the laccolithic mountain ranges, and indeed several belts of mineralization occur ringing, or immediately adjacent to laccolithic ranges, for example the Carrizo Mountains. Another interpretation is that uranium-rich volcanic ash deposits, often found in beds of the upper Chinle and Morrison formations, were eventually leached of their mineral content by groundwaters, which trickled down to plant-laden sandstones below. Whatever the original process, and there may be several processes involved in the many scattered and different mining districts, uranium deposits are numerous and rich in Navajo Country, especially in fluvial sandstones of Mesozoic age.

URANIUM FEVER IN THE SAN JUAN BASIN

As is the case with any natural resource, the value of uranium depends upon demand for the substance in the market place. Demand for uranium has historically vacillated wildly, creating economic markets only sporadically. Before its value was recognized, John Wade discovered vanadium-bearing carnotite deposits in the Carrizo Mountains, a laccolithic mountain range west of Shiprock, and staked more than three dozen mining claims between 1918 and 1920 (Chenoweth 1977, 259). Because there was no market at the time, no ore was mined. Then a local prospector, Mr. Whiteside, discovered uranium deposits in 1937 in the Grants area, as confirmed by V.C. Kelley, a professor of geology at the University of New Mexico. This discovery also fell into oblivion due to a lack of market.

With the increased demand for vanadium during World War II, the claims in the eastern Carrizo Mountains were mined by Wade, Curran, and Company, and by the Vanadium Corporation of America (VCA) during 1942–44 from exposures of the Salt Wash Member of the Morrison Formation (Chenoweth 1977). Although originally developed for their vanadium content, uranium was later taken from tailings of the mines. By 1947 it was estimated that about 12,000 tons of ore had been produced from the eastern Carrizo district, averaging about 0.27 percent U_3O_8 and 3 percent vanadium oxide.

In 1948 the United States Atomic Energy Commission (AEC) began purchasing uranium, providing incentive for exploration for the yellow ore. Uranium deposits were soon discovered in the Sanostee area south of the Carrizos, and in the Cuba–San Ysidro area along the western flank of the Nacimiento Range. Then Paddy Martinez discovered ore deposits in 1950 in the Todilto Limestone near Haystack Butte, north of the Zuni Mountains in Valencia County, New Mexico, and the prospecting rush was on in the greater Grants District. In 1951 uranium was discovered in the Morrison Formation in Poison Canyon, leading to the discovery and development of the Jackpile Mine near Laguna by Anaconda Copper Mining Company. By 1956, all surface exposures of carnotite-bearing rocks in the district had been discovered. Finding carnotite in drill cuttings taken from petroleum exploration wells, Louis Rothman began exploration by core drilling in 1955, discovering multimillion-ton ore accumulations in the Westwater Canyon Member of the Morrison Formation at what would become the Ambrosia Lake mining district (see figure 21). Lance Corporation soon discovered the larger ore body at Blackjack, and Phillips Petroleum made discoveries near Church Rock in 1958. Discoveries in the Church Rock area in 1962 by Sabre Pinon Corporation and in 1966 by Kerr-McGee led to the competitive bid lease sales by the Navajo Nation for adjacent lands in 1971, extending production eastward into the Crownpoint area where large ore bodies were developed (Chenoweth 1977). Ore production from the Westwater Canyon Member of the Morrison Formation was extended eastward from Ambrosia Lake onto the flanks of Mount Taylor in 1969 by the Fernandez Joint Venture and by Gulf Oil. Finally, the Grants mineral belt was extended eastward into the Bernabe Montaño Grant with discoveries made by Continental Oil Company (Conoco) by 1976.

Exploration extended to areas outside the Grants mineral belt and into much deeper occurrences, mostly by major oil companies, which by now dominated more expensive exploration ventures. Exxon entered into a joint agreement in 1974 with the Navajo Nation to explore 400,000 acres of tribal lands in the western San Juan Basin. Phillips Petroleum announced the discovery in 1975 of 25 million pounds of U_3O_8 north of Crownpoint, at depths of 3,000–3,500 feet, extending interest into deeper levels beyond the Grants mineral belt.

In all, between 1964 and 1977, 12,622 holes were drilled in the search for uranium, totalling more than 16 million feet of hole in the San Juan Basin area alone, according to ERDA records (Chenoweth 1977). From 1948 to 1976 more than 55 million tons of ore were mined in the San Juan Basin, containing more than 118,000 tons of U_3O_8.

ARIZONA URANIUM

Unlike the San Juan Basin, where the Morrison Formation is exposed at the surface for hundreds of miles of outcrop, occurrences of the Morrison in western regions of Navajo Country are fewer and scattered. Although there have been numerous mines that have produced from ore bodies in the Salt Wash Member of the Morrison, mostly west of the Carrizo Mountains, in the Lukachukai Mountains (see figure 21), and in adjacent areas of southeastern Utah, they have been small and relatively insignificant.

In Arizona, it has been the Shinarump Member of the Chinle Formation that has brought prosperity. The Shinarump consists of conglomeratic sandstones that occur filling isolated to broadly juxtaposed sinuous channels cut deeply into the erosional surface at the top of the Moenkopi red beds. As in Morrison sandstones, these channel-fill deposits contain significant proportions of fossil plant debris in preserved point bars. Like in the Morrison, it is the reducing local environment created within and near the fossil plant debris that has localized uranium mineralization, the uranium minerals having been precipitated from groundwater. Consequently, ore occurrences and exploration methods have been similar—only the ages of the ore bodies differ.

By far the largest and most prolific uranium mines in Arizona have been in the Monument Valley region (see figure 21). There, the Shinarump Member of the Chinle Formation is exposed along both flanks of the Monument Upwarp, extending for tens of miles along either margin of exposures of Permian rocks that dominate the crest of the uplift and cap many of the buttes and mesas within Monument Valley. The thickness of the Shinarump Member varies considerably from 10 feet to nearly 250 feet as the sporadic channels occur along the exposures.

The most important of the mines is the Monument No. 2 just south of the Utah border and directly east of Monument Valley proper. Vanadium Corporation of America leased the area in 1942. Ore deposits occur in a large scour zone, 2 miles long by 3 miles wide, where the Shinarump fills channels cut 50 feet or more into the underlying Moenkopi Formation, and in places into the Permian DeChelly Sandstone below. An inner master paleochannel, aligned in a northwesterly direction, measures 700 feet wide and 80 feet deep, where ore concentrations occur in "rods" up to 8 feet in diameter and more than 100 feet long aligned with the paleochannel trend. Logs 3 feet in diameter and 40–50 feet long, completely replaced by carnotite, including well-preserved growth rings and knots, have been high-graded from the point bar deposits. More than 500,000 tons of ore and concentrates averaging about 0.3 percent U_3O_8 were produced from this mainly open-pit mine alone, along with some vanadium, prior to being mined-out in 1967.

Production in the Monument Valley district peaked in 1955, with 14 mines operating in the area. When mining ceased in 1969, 53 properties had produced 1,362,000 tons of ore containing 8,730,000 pounds of U_3O_8 (Chenoweth and Malan 1973, 140).

Elsewhere in Arizona, uranium mining in the Cameron area (see figure 21), mostly from the Kayenta Formation and the Petrified Forest Member of the Chinle Formation, produced 289,300 tons of ore averaging 0.21 percent U_3O_8 between its discovery in 1950 and the last shipment in January 1963 (Chenoweth and Malan 1973, 141). Lesser amounts of uranium ore were produced from the Salt Wash Member of the Morrison Formation from a host of small mines in the Lukachukai and Carrizo mountains of northeastern Arizona during approximately the same time period. Navajo prospectors were responsible for the discoveries in both the Cameron and Lukachukai areas. Small amounts of uranium ore have also been produced from the upper Cretaceous Toreva Formation in the Yale Point area of northern Black Mesa, south of Kayenta.

NAVAJO INTERESTS

The Navajo Tribal Council realized the potential of income from uranium resources as early as 1949, and between 1951 and 1968 enacted 6 resolutions for the collection of royalty payments. In 1961 the Council established sliding scales for royalty payments of up to 20 percent of mine value per dry ton. The 1971 lease bids in the Grants District alone brought the Navajo Nation nearly $3 million. The joint venture between the Navajo Nation and Exxon signed in 1974 brought $6 million for exploration rights of 400,000 acres in the Shiprock, Beclabito, Red Rock, Two Gray Hills and Sanostee areas of northwestern New Mexico, and the hopes of an additional $100 million in royalties over a 10–15-year period, along with hundreds of Navajo jobs. However, the project was wrought with a great deal of local resentment from residents in those areas (Iverson 1981, 160). The nearly total decline in uranium production in Navajo Country, as well as in the rest of the United States since the late 1970s, has caused royalty payments to decline markedly—a situation most disappointing for the Navajo Nation.

POSTMORTEM

Due to market saturation, the AEC stopped buying uranium in January 1971, and ore produced after that time was purchased almost solely by the nuclear power industry. With the radical decline in the production of electricity by nuclear power plants in the 1980s and 1990s, following the moratorium on new construction since the Three Mile Island incident, the uranium mining and milling industry died. What little market is left for raw uranium is now being served almost entirely by imported supplies from cheaper foreign sources.

CHAPTER SEVENTEEN

EPILOGUE

Beauty all around me,
With it I wander.

T he legacy revealed in the record of sedimentary strata in Navajo
Country came to a close some time in the mid-Tertiary, but the
world did not end then. Erosion of the rock record began as the
entire continental interior began to rise, elevating surfaces and consequently increasing the erosional capabilities of the rivers. At first, as
stripping of the sedimentary cover began, the Colorado Plateau was
bodily tilted toward the north; Navajo Country was elevated relative to
northern parts of Utah and Colorado, and drainage carried vast quantities of newly derived sediments northward into a great lake that formed
in the Uinta Basin. As regional uplift continued, the entire Colorado
Plateau was subjected to erosional bevelling, and drainage patterns were
gradually reversed, carrying sedimentary debris toward the Gulf of California to the southwest. Exactly what changes occurred in river patterns
is uncertain and the subject of much debate; erosion has destroyed its
own history.

By Pleistocene time, some 1.5 million years ago, stream patterns
were set in stone. Rivers had carved their own destinies in deeply
entrenched canyons, carrying much of the geological records of the
province away through fixed channelways to new resting places in the
sea. Arroyos and mesas, cliffs and alcoves, formed gradually as the
destructive process progressed, until the landscape resembled the topography of the Plateau country we know today. Then glaciation of the
northern continent began, and great ice sheets blanketed the land as far
south as present-day Kansas; ice caps formed over the Rocky Mountain
ranges. The cause of the onset of glaciation is the subject of speculation,
but the results are evident. Glacial erosion stripped the north country
and the higher ranges; glacial deposits and unique erosional features testify to the sequence of events that followed.

The region that most directly affects Navajo Country is the San
Juan Mountains of southwestern Colorado, where three, and perhaps
four, episodes of glaciation interspersed with warmer and wetter periods

left morainal accumulations at the present site of Durango, Colorado (see The American Alps, University of New Mexico Press, 1992). As an episode of glaciation developed and ice advanced down the alpine valleys, the regional climate cooled; terminal moraines formed at the site of Durango to mark the southernmost terminations of ice flow. When a glacial epoch ended, ice retreated up the valleys and great outpourings of melt water rushed southward into the river valleys of Navajo Country, washing boulders and cobbles from the highlands into lower valleys as climatic conditions again warmed. With repeated periods of glacial advance and melt, subsequent flooding recurred, and the lowland valleys deepened. Complex terrace networks formed, as each succeeding flooding event cut downward into preexisting outwash deposits and left boulder accumulations perched high in its wake. A series of terraces supported by the coarse debris of glacial erosion of the San Juan Mountains resulted along the San Juan River valleys, each episode marked by a distinctive boulder-strewn terrace—higher terraces older, lower ones younger—until the present-day river level was reached during this latest interglacial epoch. The result was a series of benches along the river's course that provided habitats for plant and animal activities. The latest glacial retreat began about 15,000 years ago.

It was somewhere around that time that human habitation of the Colorado Plateau and Navajo Country occurred. Wandering hunting and gathering peoples, the Paleo-Indians of archeologists, began leaving clues as to their existence by 12,000–13,000 years ago. No doubt tracking animal herds, these ancient hunters may have followed raging, flood-engorged rivers into the ice-capped San Juan Mountains to watch the demise of the latest glacial events. The emigrants are known only by the hunting artifacts they left behind. The first recorded people were of the Clovis culture that hunted big game on the Plateau prior to 11,000 years ago, leaving behind a few distinctive projectile points known as Clovis points. A later resurgence of activity is noted by another unique style of projectile points, known as Folsom points, that represent bison hunters that wandered the region about 6,000 years ago. From these humble beginnings, the well-known reign of the Anasazi cultures began, lasting until about a.d. 1300.

But what kind of biota existed to support the vagrant Paleo-Indians? Larry Agenbroad and Jim Mead of Northern Arizona University have studied the problem for decades (Agenbroad 1990; Nelson 1990). In order to make sense from otherwise little-studied locations, they carefully excavated sedimentary debris found in the floors of caves and alcoves left in the wake of erosion, cataloging scattered bone fragments, and, most significantly, the preserved dung of ancient inhabitants. They might appropriately be called paleoscatologists. The results can be fascinating!

Hollows in the sandstone cliffs of the Plateau country have hidden the bones and excrement of animals that made these their shelters for thousands of years. Among the booty found there are the partial remains of mammalian animal communities long-since extinct from North America. The Clovis people, the first true discoverers and explorers of North America, are known to have hunted the big game—mammoths and mastodons (ancestral elephants), bison, and perhaps horses and camels. Mountain goats, tapirs, ground sloths, and musk oxen also populated the Colorado Plateau and Navajo Country. Mammoths were largely grazers of grasses, while their contemporary field companions, the giant ground sloths standing as high as 12 feet, browsed globemallow, Mormon tea, yucca, cactus and mesquite. Mammoths and sloths, horses and camels, ground sloths, tapirs and mountain goats were lost in Plateau history about 11,000 years ago as the latest of the ice ages gave way to temperate climates.

Little understood is the fact that all of these denizens originated on the North American continent, thrived here perhaps 13,000 years ago, and then disappeared as if by magic. None was found here when the first Navajos arrived in *Dinétah* more than 12,000 years later; none was found by the first Spaniards who re-explored the country in the 16th and 17th centuries. Although Indian history is almost invariably associated with the horse by most of us, those used by Native Americans were obtained in one way or another from the Spanish Fathers and their expeditions of discovery, who brought them by ship from Asia, via the Old Country, where elephants, camels and horses were commonplace.

With the extinction, or extermination, of the mammoths some 11,000 years ago, so went the Clovis people. Then came the Folsom hunters of bison around 6,000 years ago, followed by bison hunters with less specialized projectile points, the Plano culture. They thrived on the survivors of the exodus at the end of the ice ages; not only the bison, but also bears, mountain lions, wolves, antelopes and Bighorn sheep stayed around until today. But what happened to the losers—the mammoths, the sloths, the horses and camels? They must certainly have crossed the Pleistocene land bridges of the Bering Straits region into Asia, for that is where later counterparts flourished and evolved into the lineages so familiar today. But why? Was it the need to survive mass killings by the earliest Paleo-Indians? Was it the changes in climate that followed the demise of the ice ages? And what a show it must have been along the postulated land bridges—herds of animals travelling west into Asia just as droves of humans migrated eastward into North America. Wouldn't the human migrants have taken a hint that something was terribly wrong "over there"? We may never know the answers. Certainly earliest Native Americans came across from Siberia, as archeologists have insisted for decades; certainly the primitive animals must have reversed that migration at about the same time. After all, the land bridges could not have lasted long as geologic history goes.

The earliest known evidence of Navajo habitation was found in the vicinity of Gobernador and Largo canyons of *Dinétah*—Old NavajoLand—dating at about a.d. 1500. The agricultural and urban peoples now known as the Anasazi had come and gone, having left their cities and towns of the Four Corners area in about a.d. 1300, again for uncertain reasons. Navajo mythology requires that the Holy People who created *Diné* entered the land from the fourth underworld via a reed in a lake somewhere to the north; both the La Plata and San Juan mountains have been suggested as this place of origin. No previous mention of surface history is indicated, yet linguistics studies suggest that peoples speaking the Navajo language, specifically Navajos and their cousins the Apache, are closely related to present-day inhabitants of westernmost Canada and Alaska, strongly suggesting an ancestral link. Because archeologists have learned that Native Americans came into North America from Siberia across land bridges related to the Bering Straits, this linkage seems reasonable. Some ancestral Navajos may have migrated southward along the Rocky Mountain chains; some may have arrived by way of the Great Plains; mythology indicates that four clans moved eastward from southern California into *Dinétah*. Having apparently settled first in *Dinétah*—that region of north-central New Mexico and south-central Colorado east of present-day Aztec—they then spread westward into the Chuska Mountains and vicinity, perhaps because of problems caused by marauding Ute and various Plains tribes. At any rate, the Chuska Mountains and valleys, and nearby Canyon de Chelly, became the center of population of the Navajo Nation, a homeland— *Diné Bikéyah*—prior to their discovery by the Spanish explorers. It was not until the attempted round-up by Kit Carson in the disastrous winter of 1863-64 that Navajos began habitation in hide-outs in the rugged lands west of Monument Valley.

It became clear to everyone involved at Fort Sumner that *Diné* could not survive beyond the four sacred mountains—beyond *Diné Bikeyah*. Outside of these markers, set in stone by First Man and First Woman, nothing exists. The five-fingered people of the Fifth World were destined to inhabit this land of beauty. This is where the Holy People lived, where they created First Man and First Woman, where they became invisible to those of the Fifth World to serve *Diné* for all time. This is home to *Diné*.

After they created the four sacred mountains and fixed them in place with tools brought from the worlds below, First Man and First Woman created three other inner sacred mountains. It is not known exactly where one of these is located, although Gobernador Knob and Huerfano Peak are the most important; it was on the summit of Gobernador Knob that Talking God and First Man discovered the tiny baby that in only four days would mature to become Changing Woman. Some regard her as Earth itself. Perhaps the third of these sacred inner mountains is best left for *Diné* to regard as one of myriad holy places only they can fully cherish. Of only slightly lesser significance are some sacred places outsiders also recognize as being special. These include Mount Taylor and the Malpais where Monster Slayer killed Huge Monster, also known as *Big Ye'i*; Shiprock where he killed Flying Monster; Spider Rock in Canyon de Chelly where Spider Woman lived and taught *Diné* to weave and make clothing; White House ruin, also in Canyon de Chelly, where the first Nightway ceremonial was sung. These and innumerable other sacred places hold special significance in Navajo life and are of special interest to those of us outsiders who visit this ancient land because the localities are unique on Earth.

The physical—geological—origins of these special places have been discussed here, not to discredit traditional Navajo belief, but to heighten our understanding and appreciation of Navajo Country. The better we understand a remarkable region such as this, the more we appreciate its true significance. The riches found here are not only of deep aesthetic value, they have dominated Navajo life for generations untold; the natural resources have enriched us all in many ways. Now, perhaps, we can appreciate fully the underlying motives of the Holy People who knew the greater values of a seemingly barren land, who guided *Diné* to their place under the sun, who brought them back after The Long Walk to proliferate and rejoice in the land of their ancestors—where there is truly "Beauty all around me."

APPENDIX

PLACE-NAMES

A great many place-names in Navajo Country came from Spanish exploration parties who traversed the area several times in the 16th and 17th centuries. In many cases it is difficult to know who actually named which features, and in those situations where names seemed to appear from nowhere during the Spanish era, the "Spanish explorers" are credited. Among those responsible for many of the names, Coronado *circa* 1540, Juan Maria Rivera in 1765, and the friars Silvestre Velez de Escalante and Francisco Antanasio Dominguez in 1776, were the most visible; certainly Escalante is the best known because of his journals.

Derivations and history behind these place-names have been taken from numerous sources. Many are from Spanish and Navajo language dictionaries, some are from guidebooks published by the New Mexico and Four Corners geological societies, some are intuitive for lack of definitely known origins, but the conclusions are explained. Gregory's *Geology of the Navajo Country* (1917) was an important source of geographic information, as was *The Navajo Language* by Robert W. Young and William Morgan, Sr., published by the University of New Mexico Press in 1987.

ABAJO. Spanish name meaning "lower" given to the mountains of southeastern Utah by the Spanish explorers because they are lower in elevation and/or farther south than the prominent La Sal Mountains in east-central Utah. The isolated intrusive igneous (laccolithic) range is known as the Blue Mountains or The Blues by local inhabitants.

ABIQUIU, NM. Navajo=*Ha'ashgizh*, meaning "cut upward."

ACOMA PUEBLO. Name derived from the Keresan Indian words *ako* meaning "white rock," and *ma* meaning "people." The village is situated on top of a mesa composed of the white Zuni Sandstone, capped by the Dakota Sandstone. It was mentioned by Fray Marcos de Niza in 1539, and Hernando de Alvarado visited the village in 1540. Acoma was captured by the Spaniards in a siege that lasted from 21–23 January 1599. Acoma vies with the Hopi village of Oraibi in Arizona for the oldest continually occupied city in the United States.

ACOMITA, NM. Settlement west of Grants; Navajo=*Tó Łání Biyáázh*, meaning "child of many waters."

ADAH CHIJIYAHI CANYON. In Monument Valley; Navajo="where a person walked over the cliff."

AGATHLA PEAK or El Capitan. Peak composed of the neck of an ancient violently eruptive volcano (diatreme) north of Kayenta, Arizona. The name means "the place of the scraping of hides," a ritual site for communal sheep-shearing, also known as "Much Wool" (*agha*=wool; *la*=much) in Navajo language. Kit Carson is credited with applying the Spanish name El Capitan, or The Captain.

ALAMO, NM. North of Magdalena; Spanish for poplar tree; Navajo=*T'iistsoh*, meaning "big cottonwood."

ALBUQUERQUE. New Mexico's colonial governor, don Francisco Cuervo y Valdez, founded a villa in the Rio Grande valley in 1706, naming it San Francisco de Alburquerque in honor of don Francisco, Duque de Alburquerque and Viceroy of New Spain. The viceroy, fearing to offend King Philip V of Spain, renamed the villa San Felipe de Alburquerque for the king's patron saint. English-speaking people of the 19th century dropped the first "r" spelling it Albuquerque. Navajo= *Bee'eldíílahsinil* meaning "bells lie at an elevation."

ALHAMBRA ROCK. A prominent igneous (minette) dike south of Mexican Hat, UT. Apparently named for its similar appearance to La Alhambra, a Moorish Castle in southern Spain.

ANETH. Originally a trading post and now a village north of the San Juan River in extreme southeastern Utah. It has been impossible to trace the word as either a Navajo, Spanish, or family name. Regionally, the northern boundary of the Navajo Indian Reservation follows the center of the San Juan River as it occurred in 1868, but in this small corner of Utah the lands north of the river historically homesteaded by Navajo families were later added to the Reservation, and were known as the "Annex" lands. Since there has been a close association with the Spaniards and Mexicans, although not always amicable, for more than

200 years, many Spanish words with no Navajo counterparts have been included in the Navajo language. The Spanish word for annex is *anexo* (in this case the "x" is pronounced something like a soft "s"), perhaps mispronounced in Navajo as "aneth." This is the only reasonable explanation for the name that can be determined. The Navajo name is *T'áá Bííh íídii*, or a trader "he can barely make it."

ANGEL PEAK, NM. A prominent butte south of Bloomfield, NM; Navajo=*Ma'ii Dah Siké*, meaning "two coyotes seated at an elevation," probably in reference to the shape of the butte.

ANIMAS RIVER. Name given by the Spanish explorers to the river in southwestern Colorado that flows from the high San Juan Mountains east of Silverton southward through Durango and into the San Juan River near Aztec, NM, shortened from the original name *Rio de las Animas Perdido*, meaning "River of Lost Souls."

APACHE. Spanish word for thug, named for the marauding bands of Indians living in southern Arizona and New Mexico by the Spanish explorers. The name is applied to several features in the Four Corners area.

ARIZONA. From the Papago Indian word *arizonac*, meaning "place of the small spring" (*ali*=small and *shonak*=place of the spring) according to Will C. Barnes in *Arizona Place Names*. The Papago Nation is now called the Tohon O'Odham Nation. What is now southern Arizona was part of Sonora in the days of Spanish rule, and northern Arizona was part of New Mexico Territory. When the United States acquired the northern Sonoran lands with the Gadsden Purchase in 1853, the region became part of New Mexico. The name "Arizonac" was originally a station of the Saric Mission near important silver mines; the mining area was later called the District of Arizonac. The Spaniards soon dropped the "c" from the name and Padre Ortega referred to the Real de Arizona as "the town in whose district were silver mines" in 1751; the village was destroyed by Spaniards circa 1790. Arizona Territory was established by Congress on 23 February 1863, and John A. Gurley was appointed territorial governor by President Lincoln. Gurley died before travelling west of the Mississippi River and his successor, John N. Goodwin, took the oath of office at Navajo Springs in the wilds of Arizona Territory on 29 December 1863. Arizona became the last contiguous state on 14 February 1912.

AWATOBI. A ruined pueblo and spring on Black Mesa, visited by Tovar and Cardenas in 1540, destroyed in war in 1799.

AZANSOSI MESA. One of the Tsegi Mesas in Monument Valley; Navajo for "slim woman," referring to the trader's wife Mrs. John Wetherill.

AZTEC. Town in northwestern New Mexico, settled in 1878 and incorporated in 1890, named for Aztec Ruins which were erroneously believed to have been built by Indians related to the Aztecs of Mexico. The area was known as Wide House (Navajo= *Kinteel,* a name given to many large Anasazi ruins, such as Wide Ruins, AZ and Pueblo Pintado, NM) in Navajo mythology, probably for the location of the Great Kiva at Aztec Ruins, now a National Monument. The first commercial gas well in New Mexico was drilled within a half-mile of the town site; Aztec was the first community in the state to use natural gas. Because gas pressures were not regulated in the early days, many homes burned to the ground because of fluctuating pressures.

BABY ROCKS, AZ. Strangely weathered knobs and spires in the Entrada Sandstone east of Kayenta; Navajo=*Tsé'Awéé',* meaning "rock babies."

BACABI, AZ. A Hopi village; Navajo=*Tł'ohchintó,* meaning "child of wild onion spring."

BALAKAI POINT. A southeasterly promontory of Black Mesa, northwest of Ganado, AZ; Navajo="Place-of-reeds-alongside."

BEARS EARS, UT. Pair of buttes capped by the Wingate Sandstone located high atop the Monument Upwarp north of Mexican Hat; Navajo=*Shashaá.*

BEAST, THE. A volcanic neck (diatreme) in Black Creek Valley at the village of Navajo on the Defiance Plateau.

BECLABITO (or Biltabito). Trading center near the northern Arizona–New Mexico border, name meaning "water (or spring) under a rock" in Navajo language.

BEGASHIBITO CANYON. On the Shonto Plateau; Navajo *begashi*= cow, *bito*=water.

BEKIHATSO LAKE. Located in the Chinle Valley; Navajo *beek'id*=lake, *hatso'o*=large in area.

BELEN. Spanish word for Bethlehem, or "figures in Nativity scene," but colloquially means bedlam, confusion, or noise. The city south of

Albuquerque, New Mexico may have been named for the Spanish land grant of Nuestra Señora de Belen (Our Lady of Bethlehem) given to Spanish settlers by Governor Mendoza in 1740. Another possibility may be that the town site was built on a village that had been destroyed by Indians during the Pueblo Revolt of 1860, and thus the reference to bedlam.

BENNETT PEAK (Mount Bennett). Near Chaco Valley; named by the U.S. Army in 1892.

BETATAKIN RUIN. An Anasazi (Navajo word for "ancient ones" or "ancient enemies") cliff dwelling in Navajo National Monument in northeastern Arizona; name is Navajo for "ledge house," or "side hill house." Discovered by Anglos in 1909 by Byron Cummings and John Wetherill.

BIDAHOCHI BUTTE. Name of a butte and spring in the Hopi Buttes area; Navajo means "red streaks going up the sides," or "red rock slide."

BILL WILLIAMS MOUNTAIN. Near Williams, AZ, west of Flagstaff; known as "Tree-grove-slope" in Blessingway of Navajo mythology, where the stinking children were bathed before being taken to the home of Changing Woman in the west (somewhere in the Pacific Ocean).

BLACKHORSE CREEK. In the Chaco Valley, named for a Navajo headman.

BLACK KNOB. An isolated butte consisting of lava in the Painted Desert region, also known as Lava Butte.

BLACK MESA. High plateau south of Kayenta, AZ capped by rocks of Cretaceous age; Navajo *Dzilijini*=black streak mountain; so-called for its coal beds.

BLACK MOUNTAIN. Several topographic features are known by this name, the more common of which are a prominence on western Black Mesa, a volcanic neck (diatreme) east of Canyon de Chelly, and the Jemez Mountains north of Albuquerque and San Ysidro, NM.

BLACK MOUNTAIN STORE, AZ. West of Chinle; Navajo=*Tók'i Hazbi'í*, meaning "covered well."

BLACK PINNACLE. A black rock of igneous origin on the Defiance Plateau, also known as *Tsezhini*=black rock, in Navajo.

BLACK ROCK, AZ. South edge of Fort Defiance; Navajo=*Chézhiní*.

BLACK ROCK (dike between Canyon de Chelly and Canyon del Muerto). Navajo=*Chézhiní*. There are many places in Navajo Country named Black Rock.

BLANCA PEAK. In the Sangre de Cristo Range east of Alamosa, CO; the sacred mountain of the east, also called "Chief Mountain," "Horizontal-Black-Belted-Mountain," or "White-Tipped-Mountain," home of Talking God in Blessingway, the important origin legend of Navajo mythology.

BLANCO, NM. Small town located at the mouth of Canyon Largo where it enters the San Juan River east of Bloomfield, NM. Both the town and the canyon are called *Taahóóteel* in Navajo, a word meaning "valley extends down to a river."

BLANCO CANYON, NM. Navajo=*T'iistah Diiteelí*, meaning "spread among the cottonwoods."

BLANDING. Town between Monticello and Bluff in southeastern Utah established by Mormon settlers and some religious and political refugees from Mexico in 1905.

BLOOMFIELD, NM. Town south of Aztec, NM; Navajo=*Naabi'ání*, meaning "enemy cave."

BLUE CANYON. West of Fort Defiance, AZ; Navajo=*Bikooh Hodootł'izh*.

BLUEWATER, NM. Northwest of Grants; Navajo=*T'iis Ntsaa Ch'éélí*, meaning "big cottonwood."

BLUFF. A town in southeastern Utah located on the north bank of the San Juan River that was settled by Mormon missionaries of the historical "Hole-in-the-Rock" migration from Cedar City and Kanab in 1880. Named for its location beneath massive red sandstone cliffs, or bluffs.

BONITO CANYON. West of Fort Defiance on the Defiance Plateau; also known by the Spanish name Cañoncito Bonito, or "Pretty Little Canyon."

BREAD SPRING. Navajo=*Bááh Háálí*, meaning "bread flows out."

BRIDGE CANYON. Canyon west of Navajo Mountain on Lake Powell crossed by Rainbow Bridge.

BRIGHT ANGEL CREEK. A small, usually clear-water stream that drains the north rim of Grand Canyon, AZ. Named by the John Wesley Powell river expedition in 1869 because they were pleased to find clean water entering the muddy Colorado River near mid-canyon, in sharp contrast to the Dirty Devil River upstream. Since that time, several features, such as the trail and hotel, have been given the name.

BUELL PARK. An oval-shaped valley eroded from a volcanic vent, or diatreme, north of Fort Defiance, named for Major Buell of the United States Army.

BURNHAM, NM. Navajo=*T'iistsoh Sikaad*, meaning "big cottonwood."

CACHE MOUNTAIN, NM. Navajo=*Yisdá Dziil*, meaning "refuge mountain."

CABEZON. "Large head" in Spanish, an apt name for the large stone peak north of Albuquerque, NM. It is the neck of an ancient volcano exhumed by erosion; Navajo=*Gaawasóón*, the Navajo pronunciation of the Spanish name, or *Tsé Naajiin*, meaning "rock, it is black downward."

CAMERON, AZ. Trading Post on the Little Colorado River east of Grand Canyon; Navajo=*Na'ní'á Hayázhí*, meaning "little bridge" in reference to the bridge across the Little Colorado River at that location.

CAÑONCITO, NM. Spanish meaning "little pipe"; Navajo=*Tó Haji-ileehé*, meaning "water is withdrawn by a suspended bucket."

CANYON DE CHELLY. This is a strange name for the red-rock canyon in the heart of Navajo country. Numerous historians have pondered the origin of the word without success. It is pronounced "de *shay*." It seems to be a French word, applied in Navajo country, by Spaniards. Some have suggested it is a spelling corruption of the Navajo word *tsegi*, which means "rocky canyon," but that makes no sense linguistically to this writer.

I once offered a more realistic explanation: In the early 1800s, the Navajos living in the canyon grew numerous peach trees, relying heavily on the crops for food. When Kit Carson tried rounding up the Navajos to put them on a reservation, Bosque Redondo, in 1864, he tried destroying their fields to starve them out of hiding. In so doing he ordered the peach trees to be cut down in Cañon de Chelly, a Navajo stronghold since about 1700. In searching for a plausible origin of the name, I realized that the French word for peach is *pêche*, and peach tree is *pêcher*. A reasonable misunderstanding of pronunciation would easily make *pêche* or *pêcher* sound like "de-shay," or with a Spanish spelling for the sound, "de Chelly." The Frenchman who called the place "Peach Canyon" is long forgotten. Perhaps it was Antoine Leroux, a guide for the early Whipple and Sitgreaves expeditions; it certainly seems logical that the French for "Peach Canyon" became present-day "Canyon de Chelly."

Still other explanations come to light. A Navajo Park Ranger at Canyon de Chelly says that the canyon was named for a former Navajo headman who used to live there, one called *Diné Chilli'*, or "curly haired Navajo." Another, perhaps the most plausible, explanation was reported in the Navajo dictionary published by the University of New Mexico Press by Young and Morgan (1987). They offered that the Navajo name for the canyon was *Tséyi'*, a word meaning "canyon." The Spaniards, thinking that this was in itself a place name, called it Cañon de Chelly (Chelly=Spanish phonetic *Tseyi'*), hence "Canyon of Canyon."

CANYON DEL MUERTO. Tributary to Canyon de Chelly in Canyon de Chelly National Monument; name means "canyon of the dead" in Spanish. The name originated from the discovery of prehistoric Indian burials in the canyon by a Smithsonian Institution expedition led by James Stevenson in 1882. Another good reason for the name is that following Navajo raids on Pueblo Indian villages and Spanish settlements in the Rio Grande Valley, a Spanish punitive expedition led by Lt. Antonio Narbona fought an extended battle with Navajos fortified in a rock cove (now known as Massacre Cave) in the canyon, reportedly killing 115 Navajos—elders, women and children. Navajos still call the cave "where-two-fell-off" for a heroic feat by one of the women who pulled a Spanish soldier off the ledge to their deaths during the attack.

CARRIZO MOUNTAINS. Mountains in northeasternmost Arizona, west of the Lukachukai Mountains, name means "reed-grass" in Spanish. The range is called "Whirling Mountain" in Navajo mythology. It is one of the smaller laccolithic mountain clusters on the Colorado Plateau.

CARSON MESA. In Chinle Valley, named for Kit Carson.

CARSONS, NM. Also known as Huerfano, south of Farmington; Navajo=*Hanáádlí*, meaning "it flows back out."

CEBOLLETA, NM. Northeast of Grants; Spanish for "tender onion"; Navajo=*Tł'ohchin*, meaning "wild onion."

CEDAR RIDGE, AZ. Trading Post northwest of Tuba City; Navajo=*Ndeelk'id*, meaning "ridge," or *Yaaniilk'id*, meaning "slopes down and ends," with reference to the appearance of the Echo Cliffs monocline nearby.

CHA WASH. Stream and canyon on Rainbow Plateau; Navajo=*cha*, meaning "beaver."

CHACO CANYON. A National Historic Site midway between Farmington and Thoreau, NM in the heart of the San Juan Basin; large, well-preserved Anasazi surface ruins, some of the best in the Southwest, are located here. Called "Water-scattered-in," perhaps for the intricate irrigation canal system found here, in Navajo legend, it includes "Blue House" and Pueblo Bonito ("Brace-in-the-rocks") of Coyote tales.

CHAISTLA BUTTE. In Monument Valley; Navajo=beaver rincon.

CHAMA. A town in north-central New Mexico; Spanish for "barter" or "exchange"; Navajo=*Ts'í'mah* meaning "oof" as in pushing a heavy object.

CHAOL CANYON. Stream and canyon on Kaibito Plateau; Navajo= *piñon*.

CHEECHILGEETHO. Southwest of Gallup, NM; Navajo=*Chéch'iltah*, meaning "among the oaks."

CHEZHINDEZA CANYON (or Mesa). Canyon and settlement in the Carrizo Mountains; Navajo="lava."

CHILCHINBITO. Canyon and settlement on Black Mesa; Navajo= "sumac springs," or "spring in the sumac."

CHINA SPRINGS. About 5 miles north of Gallup; Navajo=*K'aatání*, meaning "many arrows or reeds."

CHINLE. Village and trading center in Navajo Country at the mouth of Canyon de Chelly and the entrance to Canyon de Chelly National Monument. The name is Navajo for "at the mouth of the canyon," also translated as "place where the water flows out of the mountain." The original trading post, operated by Naakaii Yazzie (Navajo for "Mexican Smith"), opened in 1882 from a tent.

CHINLINI CANYON. In the Carrizo Mountains; Navajo="at the mouth of the canyon"; also known as *To chinlini*.

CHURCH ROCK, NM. East of Gallup; Navajo=*Tsé 'Íi'áhí*, meaning "standing rock."

CHUSKA MOUNTAINS/CHUSKA PEAK. In northwesternmost New Mexico; Navajo=*Ch'óshgai*, meaning "white spruce"; sometimes referred to as "Rain Mountain" in Navajo legend.

CHUSKA PASS. Along old trail between Tohatchi and Crystal; Navajo=*Tsé Bii' Naayolí*, meaning "wind blows about within the rock."

CHUSKA VALLEY. Valley immediately east of the Chuska Mountains in northwestern New Mexico.

COALMINE CANYON, AZ. Navajo=*Hááhonoojí*, meaning a "rough, or jagged area."

COLORADO. Spanish word for "colored" or "red," the name applied to the river by the Spanish explorers, namely Fr. Francisco Garcés, who first saw the river in Grand Canyon. Prior to the construction of Glen Canyon Dam in 1963 the Colorado River often ran bright red through Grand Canyon due to copious amounts of red sediments it carried. Since the dam was built, the river runs red only when there are floods in the Paria or Little Colorado drainage systems that enter the Colorado below the dam. The region around the headwaters of the Colorado River (formerly known as the Grand River from Granby Lake to the confluence with the Green River) was originally called "Colorado Territory," later "Colorado State."

COMB RIDGE. A serrate ridge eroded from the upturned beds of the Glen Canyon Group (Jurassic) flanking the Comb Ridge monocline, the eastern flank of the Monument Upwarp and eastern margin of Monument Valley.

COOLIDGE, NM. Navajo=*Chíshí Nééz*, meaning "tall Chiricahua Apache."

COPPER CANYON. Stream or wash in Tsegi Mesas area west of Monument Valley; named for copper prospects found there.

CORNFIELDS. Southwest of Ganado, AZ; Navajo=*K'ii ł tsoiitah* meaning "among the rabbit brush" or "rabbit brush country."

CORTEZ. Named for Hernando Cortés, who conquered Mexico for Spain in 1519. The southwesternmost city in Colorado, Cortez is the county seat of Montezuma County, named for the native Mexican Indian chief conquered by Cortés. Navajo=*Tséyaatói*, meaning "spring under the rock."

COUNCILORS, NM. Named for the trader James Councilor; Navajo=*Bilagáana Nééz*, meaning "tall white man," referring to the trader.

CROSS CANYON, AZ. Navajo=*Béésh Dich'ízhii*, meaning "rough flintstone."

CROSSING OF THE FATHERS, UT. So named because the Spanish Fathers, looking for a northern route to the California missions, crossed the Colorado River here at Navajo Canyon on their return to Santa Fe; Navajo=*Tódáá' N'deetiin*, meaning "trail across at the brink of the water."

CROWNPOINT, NM. North of Thoreau; Navajo=*T'iists'ózí*, meaning "slender cottonwood."

CUBA, NM. Town between San Ysidro and Bloomfield, NM; Navajo=*Na'azísí To'í*, meaning "Gopher Spring."

CUBERO, NM. Navajo=*Tsék'iz Tóhí*, meaning "spring in the rock crevice."

DADASOA SPRING. In the Chuska Mountains; Navajo="corn spring."

DEBEBEKID LAKE. on Black Mesa; Navajo="sheep lake."

DEFIANCE PLATEAU. Broad plateau region west of the Chuska Mountains that forms the crest of the Defiance uplift, named for Fort Defiance.

DESHA CANYON. On the Rainbow Plateau; Navajo="curved."

DEZA POINT. A promontory, or headland, in the Chuska Mountains; Navajo="a point."

DILCON, AZ. Navajo=*Chézhin Dilkoohí*, meaning "smooth lava."

DINNEBITO SPRING. A spring and wash in the Tusayan and Painted Desert area; Navajo *Dinne* (usually *Diné*)=the Navajo people, *bito*=his spring.

DINNEHOTSO (Dennehotso). Trading post on the Navajo Indian

Reservation 26 miles east of Kayenta, AZ; name is Navajo for "people's farms." (Dictionary=*Dennihotso*, meaning "yellow streak or meadow extending up and ending.")

DIRTY DEVIL RIVER. Named by the John Wesley Powell river expedition in 1869 for the tributary stream that enters the Colorado River near Hite, Utah. Dunn rowed his wooden river boat into the mouth of the stream, and someone behind called out: "Is it a trout stream?" Dunn replied: "It's a dirty devil." The expedition, the first down the Colorado River, later found a clear stream entering near mid-Grand Canyon and named it Bright Angel Creek in contrast.

DOLORES. Spanish woman's given name applied by the Spanish explorers to the river in southwestern Colorado that drains the west side of the San Juan Mountains; later the name of the town built on the north bank of that river in 1877, incorporated in 1900. The word also refers to "sorrow" or "pain" in Spanish, probably the meaning conferred by the exploring Spaniards.

DOT KLISH CANYON. On Black Mesa; Navajo="blue."

DURANGO. A Spanish word that generally means a small town, bigger than a *pueblo* but smaller than a *ciudad*. The Chamber of Commerce version is that the name was derived from the Basque word *Urango*, meaning "water town" and that Durango was named by former territorial governor A.C. Hunt, after Durango, Mexico, which in turn is named for Durango, Spain. The town was known as Durango almost from the beginning; it was established in November 1881 by the Denver & Rio Grande Railroad after negotiations failed with nearby Animas City for the site of their railroad hub. Animas City was later merged into the northern edge of Durango and lost its identity. Navajo=*Kin ł ání*, meaning "many houses."

DULCE. Spanish word for "sweet" or "pleasant."

EGGLOFFSTEIN BUTTE. One of the Hopi Buttes; named for Frederick F.W. von Egloffstein of the Ives Expedition.

EL MORRO NATIONAL MONUMENT, NM. West of Grants, NM: Spanish probably meaning "the bluff"; Navajo=*Tséíikiin*, meaning "rock where there is water and food," or *Tsék'i Na'asdzooí*, meaning "rock upon which there is writing."

ESPEJO SPRING. On Moenkopi Plateau; named for Antonio de Espejo, Spanish explorer, 1583.

FACE ROCK, AZ.   Navajo=*Tsé Binii'í*, meaning "rock with a face."

FARMINGTON, NM.   Farming community in northwestern New Mexico turned oil-supply and drilling center in the 1950s; Navajo=*Tóta'*, meaning "between the rivers," referring to its proximity to the confluence of the Animas and San Juan rivers.

FLAGSTAFF, AZ.   City at the southwest corner of Navajo Country; Navajo=*Kin Łání*, meaning "many houses."

FLORIDA RIVER.   Spanish word meaning "flowery" or "choice."

FLUTED ROCK.   A butte bordered by columnar basalt on the Defiance Plateau; Navajo=*Dził Dah Si'ání*, meaning "mountain that sits up at an elevation"; also called "Mountain-which-holds-aloft-jewels" in Blessingway, the Navajo origin legend.

FORT DEFIANCE.   Site of an army post established in 1851 to control the raiding activities of the Navajos, and the headquarters for Kit Carson's roundup operations for "The Long Walk" to Bosque Redondo, the Fort Sumner Indian Reservation, in 1864. After the return of the Navajos to their homelands, this was the site of the first trading post in Navajo country established on 28 August 1868; Navajo=*Tséhootsooí*, meaning "meadow between the rocks."

FORT WINGATE, NM.   An army post established in 1862 and named in honor of Captain Benjamin A. Wingate under the command of General James H. Carleton at Ojo del Gallo (Spanish="Chicken Springs") near present-day Grants, NM, to suppress attacks on civilians by Navajo marauders. The fort was the headquarters, along with Fort Defiance, for the Navajo roundup by Kit Carson that led to "The Long Walk" to Fort Sumner at Bosque Redondo. Following that debacle and the return of the Navajos to New Mexico in 1868, Fort Wingate was relocated to its present site at Bear Springs (Navajo=*Shash Bitoo*), formerly known as Fort Lyon, or "old Fort Fauntleroy." Fort Wingate, built mostly of adobe and locally cut lumber, burned in July 1896 and was later deactivated in February 1911. In 1914–15 the fort became a refuge for Mexican troops and their families fleeing the attacks of Pancho Villa in Mexico. In 1918 the Ordinance Department took command of Fort Wingate, making it the largest storage depot of munitions in the world, and in 1921 the post was renamed the Fort Wingate Ordinance Reserve Depot. In 1925 the old fort facilities became a boarding school for Zuni and Navajo children, under the direction of the Bureau of Indian Affairs.

The magnificent red sandstone cliffs behind Fort Wingate were designated the type section of the Wingate Sandstone by Dutton in 1885. After considerable heated debate, it was recognized by geologists of the USGS that the cliffs not only contained the rocks generally regarded as the Wingate Sandstone across the Colorado Plateau, but also the Entrada Sandstone. The type section of the Wingate Sandstone was then restricted to only the lower cliffs, recognizing the presence of the Entrada Sandstone in the upper cliffs. After many vacillations, the age of the Wingate Sandstone is now believed to be Lower Jurassic, not Triassic as previously thought.

FOUR CORNERS. Monument where the four states Utah, Colorado, New Mexico and Arizona meet; Navajo=*Tsé 'Íi'ábí*, meaning "rock spire," apparently referring to the rock cairn that originally marked the spot.

FRUITLAND, NM. Village between Shiprock and Farmington, NM; Navajo=*Bááh Díílid*, meaning "burned bread," or *Doo 'Alk'aii*, meaning "no fat," or sometimes *Niinah Nízaad*, meaning "long upgrade."

GALLEGOS CANYON. East of Farmington, NM; Navajo=*Teeł Sikaad*, meaning a "clump of cattails."

GALLUP. The town, named for the paymaster of the Atlantic and Pacific Railroad, now known as the Santa Fe, in 1882, with the advance of the railroad. Formerly a coal distribution center, the town in northwestern New Mexico is now sometimes referred to as the Indian capital of the Southwest. It is a trading center for the Navajo who live on their reservation to the north and the Zuni Indians to the south. Navajo=*Na'nízhoozhí*, meaning "the bridge," referring to the old foot bridge over Rio Puerco (Spanish for Dirty River) opposite the Santa Fe Railroad Station in Gallup.

GALLUP HOGBACK. Navajo=*'Ał naashii Háálíni'*, meaning "springs on opposite sides."

GANADO, AZ. Location of Hubbell Trading Post National Historic Site. Spanish word for "cattle"; Navajo=*Lók'aahnteel*, meaning "wide band of reeds up at an elevation."

GANADO LAKE, AZ. Navajo=*Be'ek'id Halchíí*, meaning "red lake."

GAP, THE, AZ. North of Cameron; a pass through the Echo Cliffs monocline; Navajo=*Tsinaabaas Habitiin*, meaning "wagon trail up out."

GARNET RIDGE. A volcanic neck (diatreme) south of the San Juan River along Comb Ridge; a well-known source for garnets.

GOBERNADOR KNOB. Spanish word meaning "governor"; a sacred mountain known in Navajo legend as "Spruce Hill" (also "Spruce Mountain" or "Mount Lookout," sometimes "Precious Fabrics Mountain" or "Hard Goods Mountain")—the model for the pointed-upward hogan (male hogan) to be used in the Blessingway ritual, birthplace of Changing Woman who was found by Talking God and raised by First Man and First Woman, put in charge of Blessingway by the Holy People.

GOODRIDGE BRIDGE. Bridge across the San Juan River at present site of Mexican Hat village; named for E.L. Goodridge, prospector in the late 1890s and early 1900s and discoverer of the Mexican Hat (San Juan) Oil Field in 1907.

GOTHIC MESA. An oddly eroded plateau south of the San Juan River, overlying part of the "greater Aneth" oil field; named by Macomb in 1859.

GOULDINGS. Trading Post, Motel, and Navajo boarding school near Monument Valley in northeastern Arizona. Harry Goulding came to Monument Valley in 1923 and established the trading post for Navajo trade, building the stone structure in 1927 and adding the first tourist quarters in 1928–29. He made Monument Valley famous when in the 1930s he coaxed Hollywood film producers into making western movies here, mostly starring John Wayne; now a favorite location for television commercials.

GRANADA. Trading post and mission in Navajo country near the intersection of Arizona Highways 63 and 264 west of Window Rock, AZ. Name means "cattle" in Spanish. The Hubbell Trading Post, established here in 1876 by John Lorenzo Hubbell, was designated a National Historic Site in February 1967 to preserve a realistic picture of the old Navajo trading days.

GRAND CANYON, AZ. Navajo=*Bidáá Ha'azt'i'*, meaning "something extends in a slender line up to its edge," referring to the railroad tracks.

GRAND FALLS, AZ. Waterfall in the Little Colorado River near Leupp; Navajo=*'Adahiilíní*, meaning "waterfall."

GRANTS, NEW MEXICO. This favorite off-reservation trading center for the Navajos, and mining, lumbering, and ranching center, is named for the Grant brothers (Augustus, Lewis, and John) who were railroad construction contractors and maintained a camp for railroad workers here known as Grants Camp. The site was later a coaling station for the AT&SF Railroad known as Grants Station. Navajo=*Naatooh Sik'ai'í*, meaning "fork-legged Isleta," referring to a former Isleta Indian prostitute of Grants.

GREASEWOOD SPRING. Several springs are known by this name; Navajo=*Duwuz* or *Hibito*, meaning "spring among the greasewood."

GREAT PLAINS. The Midcontinent, USA; Navajo=*Halgai Hóteelígíí*, meaning "broad white area."

GROANING LAKE. A small, usually dry lake 8 miles southwest of Chinle, AZ; Navajo=*Be'ek'id Di'nini.*

HAMBLIN CREEK. Wash in the Painted Desert region, named for Jacob Hamblin, a Mormon guide who assisted Major John Wesley Powell in the mid-1800s.

HANO, AZ. A village in Hopi Country on southern Black Mesa; Navajo=*Naashashí*, meaning "bear enemies."

HARD ROCK, AZ. North of Oraibi; Navajo=*Tsé Dildó'í.*

HASBIDITO VALLEY. In the Chuska Mountains; Navajo="turtle dove."

HOGBACK, NM. East of Shiprock at the Navajo Indian Reservation boundary; Navajo=*Tsétaak'á*, meaning "rock ledge slants into water," referring to the steeply dipping sandstones of the Mesaverde Group (Cretaceous) along the Hogback monocline—the sandstones dip southeastward into the San Juan River.

HOLBROOK, AZ. Town about midway between Gallup, NM, and Flagstaff, AZ; Navajo=*T'iisyaakin*, meaning "house under the cottonwood."

HOPI BUTTES. A region of many igneous extrusive vents (diatremes) in the southeastern and adjacent parts of the Hopi Indian Reservation.

HOSKINNINI MESA. A prominent mesa west of Monument Valley in the Tsegi Mesas area; named for a Navajo headman who hid in this area with his followers from capture by Kit Carson and "The Long Walk"; Navajo=*Hashké*, meaning "mean warrior," *Neiniihii*, meaning "he distributed them (sheep) with angry resistance." He is said to have hid from the Kit Carson roundup in the very rough and isolated region west of Monument Valley and, believing that other Navajos would soon return from Bosque Redondo, he and his followers built up their herds for redistribution to the returning Navajos. When they refused his contribution of large parts of his herds, "he would not take no for an answer," hence the name. The mesa is the type section for the Hoskinnini Member of the Moenkopi Formation, Triassic.

HOSTA BUTTE. Navajo=*'Ak'i Dah Nást'áni*, meaning "the one (mountain) that sits on top of another."

HOTEVILLA, AZ. A Hopi village; Navajo=*Tł 'ohchintó*, meaning "onion spring."

HOUCK, AZ. Town on Interstate Highway 40 between Lupton and Chambers; Navajo=*Ma'iito'í*, meaning "Coyote Spring."

HOVENWEEP NATIONAL MONUMENT. Reserved for scattered, mostly surface Anasazi Indian ruins in south western Colorado on 2 May 1923; name is Ute Indian for "deserted valley."

HOWELL MESA, AZ. Navajo=*Tsin Bił Dah 'Azkání*, meaning "timbered mesa."

HUERFANO MOUNTAIN. Spanish word for a male orphan, a broad mesa south of Farmington, near Huerfano, in the San Juan Basin; called "Mountain-around-which-moving-was-done," sometimes "Jewel Mountain" in Navajo mythology. The place designated as the model for dome-shaped hogans in Blessingway, where Injuryway and other chantways are to be sung; hogan of young Changing Woman was at its base on the west side; personified by Pollen Boy and Cornbeetle Girl; home of Coyote, its official protector.

HUNTERS POINT, AZ. Between Lupton and Window Rock: Navajo=*Tsé Náshchii'*, meaning "red rock in a circle," or *Tsé Naachii'*, meaning "red rock coming down," both descriptive names referring to the monoclinal drag fold along a fault at that location.

ICE CAVES. Caves in the Malpais (lava flows) south of Grants, NM; Navajo=*Dibé Hooghan*, meaning "sheep house."

IGNACIO, CO. Village southeast of Durango, headquarters for the Southern Ute Indian Tribe; named for a Ute chief; Navajo=*Bíina*, meaning "pine."

INDIAN WELLS, AZ. North of Holbrook; Navajo=*Tó Hahadleeh*, meaning "water is drawn out one quantity (bucket?) after another."

INSCRIPTION HOUSE, AZ. Anasazi ruin in Navajo National Monument near Shonto; Navajo=*Ts'ah Bii'kin*, meaning "house in the sagebrush."

ISLETA PUEBLO. Spanish name for a village built on a "little island" surrounded by the Rio Grande flood plain. The original pueblo was located on the present-day site when Coronado visited this area in 1540. Raids by Plains Indians caused Pueblo Indians living east of the Manzano Mountains to move here around 1675. The Isleta Pueblo did not participate in the Pueblo Revolt against the Spaniards in 1680, but Governor Otermin captured 400 to 500 prisoners from Isleta in 1681. Others living in the pueblo escaped to Hopi country in Arizona, returning in 1716 with Hopi relatives. Residents of Acoma and Laguna moved to Isleta in the early 1800s as a result of droughts and religious problems. Thus, Isleta is a community of a variety of migrants. Navajo=*Naatoohó*, meaning "the river enemy," referring to the Isleta Indians.

IYANBITO. Settlement east of Gallup, NM; Navajo meaning "buffalo spring."

JACOB'S WELL (15 miles south of Sanders, AZ): Navajo='*Ahoyoolts'i∤*, meaning "hole that increases in size."

JEMEZ PUEBLO, NM. Pueblo and settlement north of San Ysidro; Navajo=*Ma'ii Deeshgiizh*, meaning "Coyote Pass."

JOSEPH CITY, AZ. Town west of Holbrook; Navajo=*Náyaaseesí*, meaning "one (former trader) with a wart under the eye."

JOYITA. Spanish for "small jewel"; low hills near Socorro, NM.

KAIBITO. A spring in the Hopi Buttes country, sometimes called Comar Spring; Navajo="willows at a spring."

KAYENTA. Trading center on the Navajo Indian Reservation in northeastern Arizona, originally a trading post operated by John and Louisa Wetherill founded in 1910. The location was originally called Todanestya or *Tó Dínéeshzhee'* (Navajo meaning "where water runs like fingers out of a hill"), later changed to Kayenta, an Anglo corruption of the Navajo word *Tyende* ("where the animals bog down"). John

Wetherill made significant archeological discoveries of Mesa Verde, now a National Park, and Betatakin, Keet Seel, and Inscription House ruins, now in Navajo National Monument, and was the first white man to see Rainbow Bridge. Kayenta, pronounced something like a sloppy "kanta" by local people, was once known as "the farthest place from anything" in the United States.

KEAMS CANYON. Settlement and canyon on Black Mesa; named for Thomas U. Keams who made the first settlement here in Hopi Country; Navajo=*Lók'a'deeshjin*, meaning "reeds extend along black."

KEET SEEL. The name given Anasazi cliff dwellings in Navajo National Monument in the Tsegi Mesas area; Navajo=*Kits'iilí*, meaning "shattered house."

KINLICHEE, AZ. A small Anasazi ruin called "red house."

KIT CARSON CAVE, NM. East of Gallup; Navajo=*Tsé'áhálzhiní*, meaning "black cave."

KLAGETOH, AZ. Between Ganado and Chambers, AZ. Navajo word meaning "water in the ground" or "spring."

LADRON MOUNTAINS. Spanish word for "thief" or "robber."

LAGUNA, NM. Settlement west of Grants; Navajo=*Tó Łání*, meaning "many waters."

LA JARA, NM. Spanish name meaning "the rock-rose"; Navajo=*K'ai Ch'ínee ltł'ó*, meaning "tangle of willows extending out."

LA JUNTA. Spanish word for "board," as in a governing board.

LAKE VALLEY, NM. Navajo=*Be'ek'id Halgaii*, meaning "lake where the area is white."

LARGO CANYON, NM. A large canyon, tributary to the San Juan River east of Aztec, NM in *Dinétah* (Old Navajoland); Navajo='*Ahi-dazdiigaii*, meaning "treeless areas come together," or *Tsííd Bii' Tó*, meaning "spring in the embers."

LA SAL. Spanish word for "salt," the name given to the mountain range in east-central Utah by the Spanish explorers. The name is most appropriate as the mountains were formed by igneous intrusions into thick salt-thickened flowage structures in middle Tertiary time. How the Spaniards knew the relationships of the mountains to the completely subsurface occurrence of salt is a mystery, but there have been unconfirmed reports of salt water springs in these mountains. Also, salt water springs along the Dolores River in Paradox Valley south of the range could have been the tip-off.

LA PLATA. Spanish word for "silver," named by Spanish explorers led by Juan Maria Rivera in 1765 who found silver minerals in the mountain range. La Plata County, formed in 1874 with Parrot City as the county seat (now Durango is the county seat), was named for the mineral-rich laccolithic range.

LA PLATA MOUNTAIN (Mount Hesperus). Sacred mountain of the north; Navajo=*Bááshzhinii Dziil* or *Dibé Ntsaa*, meaning Jet Mountain; also known as Big Sheep Mountain. Some believe that the location of the legendary Emergence Hole (*Hajíínái*) is in the La Plata Mountains.

LEUPP, AZ. Northwest of Winslow; Navajo=*Tsiizizii*, meaning "scalplock" (hair root?).

LOHALI POINT. Prominent point on Black Mesa; Navajo=*Łóó*, meaning "spring where there are fish."

LONE MOUNTAIN, AZ. Navajo=*T'áá Sahdii Dah*, meaning "the separate mesa."

LONG LAKE. South of Washington Pass in the Chuska Mountains; Navajo=*Be'ek'id Hóneezi*.

LOS GIGANTES, AZ. Spanish for "giants"; Navajo=*Tsé Ch'ídeelzhah*, meaning "rocks jut out."

LUCERO MOUNTAINS. Spanish word meaning "Morning star," or "splendor."

LUKACHUKAI. Mountains and settlement in northwesternmost NM north of Chuska Mountains, name translated from the Navajo language variously as "place of slender reeds," "patches of white reeds," or "Reed-extends-out-white," as from a canyon.

LUPTON, AZ. Southwest of Gallup; Navajo=*Tsé Dijoolí*, meaning "sitting rock."

LYBROOK, NM. Southeast of Bloomfield; Navajo=*Tó Náálíní*, meaning "water flows downward."

MANCOS. River, Valley, and city in southwestern Colorado. Spanish word for those who are "armless," "defective," or "faulty." The reason for the Spanish explorers to apply this term to the Mancos River is probably because the river has no major tributaries. Known as Slim Water Canyon (*Tó Nts'ósíkooh*), this is where Coyote courted Tingling Maiden in Navajo mythology.

MANUELITO, NM. Spring and plateau in Chuska Valley; named for respected Navajo headman; Navajo=*Kin Hóchxó'í*, meaning "ugly house," for a nearby Anasazi ruin.

MANY FARMS, AZ. Navajo=*Dá'ák'eh Haláni*, meaning, not surprisingly, "many fields."

MANZANO. Spanish word for "apple trees" which grew in two orchards near the village on the east slope of the Manzano Mountains, southeast of Albuquerque, New Mexico. The trees were originally believed to have been planted during the Spanish missionary period before 1676, but dating of the growth rings of the trees has established that they were planted no longer ago than 1800.

MARIANO LAKE, NM. Navajo=*Be'ek'id Hóteelí*, meaning "broad lake."

MARSH PASS. Near Kayenta, AZ; Navajo=*Bitát'ah Dzígai*, meaning "white streak on the ledge," or *Tsé Yík'áán*, meaning "hogback."

MATTHEW'S PEAK, AZ. Navajo=*Chézhin Náshjiní*, meaning "lava with a black band around it," or *Tsé Binááyołí*, meaning "rock around which the wind blows."

MERIDIAN BUTTE. In Monument Valley, a prominent butte near the 110th meridian.

MESA VERDE. A descriptive term used by the Spanish explorers to describe the "green tableland" between Mancos and Cortez in southwestern Colorado. The high wilderness region was the homelands of Ute Indians in the mid-1800s, and white settlers avoided the region. The first discovery of Anasazi ruins was reported by W.H. Jackson, photographer for the United States Geological and Geographical Survey, who photographed the region in 1874. However, the discovery of Cliff Palace by two cowboys, Richard Wetherill and Charlie Mason, in December 1888 brought notoriety to the region. It was made a National Park on 29 June 1906 to display and preserve Anasazi Indian artifacts. Navajo=*Gad Deelzha*, meaning "jagged juniper mountain," or

Nóóda Dził, meaning Ute Mountain. Known as Rock Point Mesa in Navajo mythology, probably for the prominent Point Lookout on the north of Mesa Verde, where a trail led Coyote on numerous trips between his family home at Shiprock (Winged Rock) and the Mancos River (Slim Water Canyon) during his courtship of Tingling Maiden.

MESITA, NM. Southeast of Laguna; Spanish for "little mesa"; Navajo=*Tsé Ch'échii'*, meaning "horizontal red rocks."

METEOR, AZ. Navajo=*'Adah Hosh Láni*, meaning "many cacti down from a height."

MEXICAN WATER. Trading Post on the Navajo Indian Reservation in northeastern Arizona, Navajo=*Naakaii-Tó*, meaning "Mexican water" as the freshwater springs emanating from the Navajo Sandstone have been used by travellers for decades, and perhaps centuries.

MEXICAN HAT. Village in southeastern Utah located on the San Juan River north of Monument Valley, named for a monolith near the town that is shaped like a conical peak, capped by a very large, round balanced rock, giving the appearance of a dozing Mexican wearing a sombrero. The town was originally known as Goodridge, named for the turn-of-the-century prospector who discovered shallow oil at the site in 1908. The Mexican Hat town site was then located about 1.5 miles to the north; Mexican Hat was abandoned in 1930, and the name was transferred to the present site. The Mexican Hat Field still produces oil in barely economical quantities; miniature pump-jacks can be seen scattered around the countryside, in and near the town. Navajo=*Naakaii Ch'ah*, meaning "Mexican Hat."

MITTEN BUTTE(S). Named for their mittenlike shape, a pair of buttes that typify Monument Valley; buttes consist of slopes of Organ Rock Shale underlying cliffs of DeChelly Sandstone (Permian), capped by the Hoskinnini Member of the Moenkopi Formation (Triassic).

MOENAVE. Settlement and spring on the Kaibito Plateau west of Tuba City, AZ.

MOENKOPI. Settlement and wash south of Tuba City, AZ; a Hopi word for running water. Navajo=*Naak'a' K'éédílyéhé*, meaning "where cotton is raised," or sometimes *'Oozéí Hayázhí*, meaning "Little Oraibi." Also used as the type section of the Moenkopi Formation (Triassic).

MOLAS LAKE. Scenic alpine lake south of Silverton, Colorado in the heart of the San Juan Mountains. The name is Spanish for "moles," actually present are marmots that commonly dot the lakeshore areas with burrow mounds. (Correctly pronounced "MOLE-us," not "mole-ASS.") This is perhaps the "Emergence Place," mentioned in Navajo origin stories; the place where the Holy People emerged from the underworld to become Earth Surface People in Blessingway, the Navajo origin legend.

MONTICELLO. Town in southeastern Utah established by Mormon settlers from Bluff, Utah in 1886.

MONUMENT CANYON. (in Canyon de Chelly). Navajo=*Dzaanééz Ch'íbítiin*, meaning "where the mules exit."

MONUMENT VALLEY. Area containing numerous red sandstone monoliths sitting astride the Monument Upwarp on the Arizona-Utah border between Kayenta and Mexican Hat; Navajo=*Tsé Bii' Ndzisgaii*, meaning "treeless areas (clearings) among the rocks."

MOQUI BUTTES. (Also Hopi Buttes). An area of numerous volcanic vents (diatremes) in the southeastern Hopi Country in Arizona; Navajo=*Dibé Dah Sitíní*, meaning "reclining sheep up at an elevation."

MOUNT HESPERUS. Highest peak in the La Plata Mountains, a laccolithic range west of Durango in southwestern Colorado. It is the sacred mountain of the north, or "Big Sheep Mountain," in Navajo myth; place of dark clouds and jet; home of White Corn Boy and Yellow Corn Girl.

MOUNT TAYLOR. Coronado passed by Mount Taylor in 1540 on his way to Acoma and mentioned the volcanic features in the area. The ancient volcanic cone was known variously as Sierra San Mateo and Cebolleta (tender onion) to the Spaniards, and *Tsoodził*, the sacred mountain guarding the southern boundary of Navajo country, to the Navajos. Also known as "mountain-which-the-wind-strikes" or "Tongue Mountain," home of Boy-who-returns-with-a-single-turquoise and Girl-who-returns-with-a-single-(ear of)-corn, or "Turquoise Mountain" in Navajo mythology. Simpson, working on transcontinental railroad rights-of-way in 1850, named the mountain after the president.

MOUNT THOMAS. Also known as Mogollon Baldy Peak in the White Mountains, AZ. One of the Holy Places of Navajo mythology, especially in Blessingway, where it is called "Yucca Mountain."

NACIMIENTO MOUNTAINS. A north-trending mountain range north of Albuquerque and San Ysidro, NM. The upland originated as a fault block in Precambrian time and was actively high until Permian time. Name is the Spanish word for "birth," "nativity," or "origin."

NASCHITTI, NM. Settlement north of Gallup on U.S. Highway 666; Navajo word meaning "badger."

NATURAL BRIDGE, AZ. Southwest of Fort Defiance, and several other natural bridges; Navajo=*Tsé Naní'áhí*.

NAVAJO. Name given to the local Indians living in the general region of the present-day San Juan Basin by the Spanish explorers. The origin of the name is in doubt, for it has several meanings. Some historians believe the name was derived from a Tewa-speaking Pueblo Indian place-name *navaju*, which is believed to have meant "large area of cultivated lands." However, the modern Spanish word *navajo* means either "plain" or "stupid." We prefer to think that the arrogant Spanish explorers referred to the Indians as "plain," or unsophisticated people. There are at least two settlements of this name in Navajo Country—one north of Fort Defiance at "The Beast," and another on Interstate Highway 40 west of Chambers, AZ.

NAVAJO BRIDGE, AZ. Bridge across Marble Canyon built in 1928 to replace Lees Ferry, a historic crossing of the Colorado River; Navajo=*Na'ní'á Hatsoh*, meaning "big span."

NAVAJO CANYON. Navajo=*Tsékooh Niitsí'ii*, meaning "cheek canyon."

NAVAJO MOUNTAIN. High (elevation 10,384 feet) prominent landmark east of Lake Powell in northeastern Arizona, a structural dome formed by the intrusion of igneous rocks. It is called *Not is Ahn* or "place of the enemies" by Navajos. Navajo Mountain was designated the headrest for Mountain Woman, a male member of the Holy People (Agathla Peak was his cane), in Blessingway, the central origin ritual in Navajo mythology; also known as *Tádídíí Dziil* meaning "Pollen Boy."

NAVAJO PEAK. Near Chromo, CO; Navajo=*Dziɬ Binii'*, Ligaii meaning "white faced mountain."

NAZLINI. A settlement and prominent canyon south of Canyon de Chelly in Navajo country; name is Navajo for "place where it (wash) makes a turn flowing," also translated as "runs crooked."

NEWCOMB, NM. Also known as Nava; Navajo=*Bis Deez'áhí*, meaning "adobe extends."

NIKAHOSHI SPRING. In the Chuska Mountains; Navajo="one eye," named for a Navajo man.

NOKAI. Navajo="Mexican" and Nokaito="Mexican water"; names given to numerous streams and canyons in Navajo Country.

NUTRIA, NM. Spanish meaning "otter"; Navajo=*Tsé Dijíhí*, meaning "rock starts to extend black."

OAK SPRINGS CANYON, AZ. South of Window Rock; Navajo=*Tsét ł 'áán Ndíshchí'í*, meaning "pines in a crescent around the canyon's edge."

OLJETO. A village and trading post west of Monument Valley; Navajo="moonlight water."

OJO ENCINO, NM. Spanish for "eye adorned"; Navajo=*Chéch'il Dah Ł chí'í*, meaning "oak up at an elevation"; also *Chéch'iizh Bii' Tó*, meaning "spring in the rough rock."

ORAIBI. Largest and perhaps the oldest of the Hopi pueblos on Black Mesa; Hopi="place of the rock"; Navajo=*'Oozéí*.

PAINTED DESERT. Region lying between Black Mesa and the Kaibab Uplift where varicolored rocks of the Chinle Formation (Triassic) are well exposed; name from Ives in 1861. Navajo=*Halchíítah*, meaning "among the red areas."

PARADOX. Name of a village and long valley in west-central Colorado. The valley, which extends into easternmost Utah, is the collapsed top of a salt flowage structure, known as a salt anticline; a deep well drilled into the valley penetrated nearly 15,000 feet of salt and closely related sedimentary rock. The Dolores River crosses the structurally controlled valley at nearly right angles, rather than flowing DOWN the valley as other rivers do. The course of the river had been established on a surface far above the present-day land level, and regional erosion removed the overburden from the salt structure. When the river bed cut down to the top of the northwest-trending elongate structure, the river was hopelessly trapped in its canyon and cut ACROSS the structure. Subsequent erosion and collapse of the salt anticline formed the present-day valley. Seeing this as a most unusual situation, pioneers in the region named this "Paradox Valley" and the village also took that name. The usually deeply buried salt strata were named the "Paradox Formation," and the ancient basin in which the salt was deposited was named the "Paradox Basin" for this location. A popular T-shirt available in the area asks the burning question "Where in the hell is Paradox?" answered on the back as "Right next to Bedrock."

PARIA RIVER. Tributary to the Colorado River entering just below Lees Ferry, AZ. The word is Spanish for "pariah," or "outcast."

PAQUATE, NM. East of Grants; Navajo=*K'ish Ch'ínít'i'*, meaning "they extend out horizontally in a line."

PETRIFIED FOREST, AZ. A National Park east of Holbrook, AZ where an abundance of petrified fossil wood has weathered out of exposures of the Chinle Formation (Triassic). Navajo=*Sahdiibisí*, meaning "the place where there is a lone adobe monolith," or *Tsé Nástánii*, meaning "stone logs."

PIEDRA RIVER. The Spanish word for "stone," or "gravel," perhaps best translated as "rocky river."

PINEDALE, NM. Settlement northeast of Gallup; Navajo=*Tó Bééhwíisganí*, meaning "area that is dry around the water."

PINE RIVER. Originally named the Rio de los Piños by the Spanish explorers, now known by the English translation.

PINE SPRINGS, AZ. Navajo=*T'iis 'Íí'áhí*, meaning "standing cottonwood."

PIÑON, AZ. Navajo=*Be'ek'id Baa 'Ahoodzání*, meaning "pond with a hole in it," referring to a drilled well,

PIUTE FARMS. Site of former marina on Lake Powell, west of Monument Valley; Navajo=*Báyóodzin Bikéyah*, meaning "Piute country."

PUEBLO BONITO, NM. Spanish for "pretty town"; Navajo=*Tsé Bíyah 'Anii'áhí*, meaning "rock under which something extends supporting it," referring to an Anasazi rock wall built to support a rock behind a ruin. (The rock fell in March 1941.)

QUARTZITE CANYON. Canyon west of Fort Defiance where quartzite beds of Precambrian age are exposed.

RAINBOW BRIDGE. A large natural bridge, the central attraction of Rainbow Bridge National Monument west of Navajo Mountain, now on Lake Powell; Navajo=*Nonnezoshi*, meaning "great arch," or *Tsé'naa Na'ní'áhí*, meaning "span across."

RAMAH, NM. Southeast of Gallup; Navajo=*Tł 'ohchiní*, meaning "wild onion."

RATON SPRINGS, NM. In north-central New Mexico; Spanish for "rat"; Navajo=*Tó Dích'íí*, meaning "bitter water."

RATTLESNAKE, NM. Small settlement west of Shiprock at site of former Rattlesnake Oil Field; Navajo=*Siláo Habitiin*, meaning "soldiers' ascending trail."

RED LAKE, AZ. Navajo=*Be'ek'id Halchíí*, meaning "lake where the area is red," or Red Lake.

REDONDO. Spanish for "round."

RED ROCK MESA, AZ. Navajo=*Tse' Łichíí Sikaadí*, "red rock spreading flat."

REDROCK VALLEY. A region north of the Chuska Mountains, named for the color of its exposed rocks of Mesozoic age.

RIO GRANDE. Spanish term for "Grand River," often misused in English as the Rio Grande River, meaning "River Large River." Navajo=*Tooh Ba'ááá*, meaning "female river," probably because of its slow-moving waters and low gradient.

RIO PUERCO. Spanish name meaning "Dirty River" or "Pig River," an apt descriptive name for the central New Mexico intermittent stream.

RIVER JUNCTION. (also "Parallel Streams"). Probably the confluence of the Pine *(Los Piños)* and San Juan Rivers; Navajo="Waters-flowing-together." Important mythical locality in Blessingway, the important Navajo origin legend.

ROCK SPRINGS, NM. Navajo=*Chéch'ízhí*, meaning "rough rock."

ROCKY POINT, NM. West of Gallup; Navajo=*Chézhin Ditłooí*, meaning "fuzzy traprock."

ROOF BUTTE. Highest peak in the Lukachukai Mountains north of Fort Defiance; Navajo=*Dził Dah Neeztínii*, meaning "mountain that lay down."

ROUGH ROCK, AZ. Navajo same as Rock Springs above.

ROUND ROCK, AZ. Northwest of Lukachukai; Navajo=*Tsé Nikání*.

SACRED MOUNTAINS. Four external sacred mountains designate the limits of Navajo Country *(Diné Bikéyah)*, as specified in Blessingway, the significant origin legend in Navajo mythology: the sacred mountain of the East is Blanca Peak east of Alamosa, CO; of the south is Mount Taylor near Grants, NM; of the west is San Francisco Peak (Mount Humphries) north of Flagstaff, AZ; of the north is Hesperus Peak, in the La Plata Range west of Durango, CO. Two inner sacred mountains were also formed at Gobernador Knob and Huerfano Mountain in the San Juan Basin of northwestern New Mexico. In practice, all mountains may be considered sacred.

ST. JOHNS, AZ. Navajo=*Chézhin Deez' áhí*, meaning "lava point" or "ridge."

ST. MICHAELS, AZ. Trading center west of Window Rock; Navajo=*Ts'ihootso*, meaning "meadow extends out horizontally."

SALT POINT, NM. A ruin at the junction of Canyons Largo and Blanco east of Aztec, NM; Navajo=*'Ashiih Náá'á*, meaning "salt extends downward."

SANDERS, AZ. Navajo= *Łich'íí' Deez'áhí*, meaning "red bluff."

SANDERS, NM. Navajo=*Chííh Tó*, meaning "red ochre water (spring)."

SANDIA. Spanish word for "watermelon," named for the sliced watermelon-like appearance of the Sandia Mountains.

SAND SPRINGS, AZ. On Ward Terrace northeast of Flagstaff, AZ. Navajo=*Séí Bii' Tóhí*, meaning "spring in the sand."

SAN FELIPE PUEBLO. Navajo=*Dibé Łizhiní*, meaning "black sheep," or sometimes *Séí Bee Hooghan*, meaning "sandy home," or *Tsédáá' Kin*, meaning "house on the edge of a cliff."

SAN FRANCISCO PEAKS. A region of recent volcanic activity north of Flagstaff, AZ. Mount Humphries, the highest summit, is the Navajo sacred mountain of the west, *Dook'o oosłííd*, called "light-shines-from-it," or "it has never melted and runoff from it," or "Faultless Mountain," or "Abaloné Mountain"; home of Calling God, in Navajo mythology.

SAN JUAN. Name given to the river and mountain range in southwestern Colorado by the Spanish explorers, probably Juan Maria

Rivera in 1765, meaning "Saint John." Navajo=*Tooh Biká*, meaning "male river," probably in reference to its fast-moving water and rapids.

SANOSTEE, NM. South of Shiprock. Name is from the Navajo *Tsé 'A ł náozt'í'í*, meaning "layered rocks"; Navajo name used for the locality, however, is *Tó Yaagaii*, meaning "artesian (flowing) well."

SAN RAFAEL, NM. South of Grants; Navajo=*Tó Sido*, meaning "hot water."

SANTA CLARA PUEBLO. Navajo=*'Anaashashí*, meaning "the bear enemies."

SANTA FE. Name given to the Spanish government city in north-central New Mexico in the 17th century, the name meaning "Saint of Faith." The city has since been the site of the Spanish, Mexican, United States Military, and State capitols. Navajo=*Yootó*, meaning "bead-water."

SANTA DOMINGO PUEBLO, NM. Navajo=*Tó Hajiiloh*, meaning "people draw up water."

SAWMILL, AZ. Former site of Navajo-owned sawmill west of Fort Defiance; Navajo=*Ni'iijíhí*, meaning "where sawing is done."

SEVEN LAKES, NM. Northeast of Crownpoint; Navajo=*Tsosts'id Be'ak'id*, meaning, you guessed it, "seven lakes."

SHEEP HILL. East of Flagstaff, AZ; Navajo=*Dibé Dah Shijé'é*, meaning "reclining sheep up at an elevation."

SHEEP SPRINGS, NM. North of Gallup; Navajo=*Tó Haltsooí*, meaning "water in a meadow."

SHIPROCK. Name given to the prominent monolith in northeasternmost New Mexico and the nearby town for the sailing-shiplike appearance of the rock, which rises 1,800 feet from a nearly flat and barren plain. The monolith is the remains of a diatreme, the preserved neck of a blowout volcano. Shiprock is considered to be a holy place by Navajo Indians, who call it *Tsé Bit'a'í*, "the winged rock", where Monster Slayer killed the Flying Monster, and home of Coyote's family.

SHIPROCK, NM. Town in northwesternmost New Mexico; Navajo=*Naat'áanii Nééz*, meaning "tall boss," referring to former Superintendent William T. Shelton; also called *Tooh*, meaning "river."

SHONTO. Name given to springs, a stream, and a plateau between Kayenta and Page, AZ, in the Tsegi Mesas area; Navajo=*Sháá'tóhí*, meaning "water-on-the-sunny-side-of-a-rock-wall. "

SHUNGOPOVI. A Hopi village on Black Mesa, AZ; Navajo=*Kin Názt'i'*, meaning "house strung in a circle."

SICHOMOVI. One of the Hopi villages in Arizona; Navajo=*'Ayahkin*, meaning "underground house."

SILVERTON. Name of the village in the heart of the San Juan Mountains that was the mining center mostly for silver in the late 1800s, although modern sporadic mining operations in the area are mainly for gold. The first year-round settlers arrived in about 1874. The initial settlement was named "Baker's Park" for Charles Baker who led a party of prospectors into the area in 1860–61. The party was harassed by Ute Indians and heavy snows and escaped southward to near the site of Baker's Bridge north of Durango. This was the first group of whites to have visited the site of present-day Durango. Baker was later killed by the Indians.

SNOWFLAKE, AZ. Small town on the Mogollon Rim; Navajo=*Tó Dił hił Biih Yílí*, meaning "where it flows into the dark water."

SOCORRO. Spanish for "HELP!" You should have seen Socorro, New Mexico in the early days.

SONSELA BUTTES. Buttes capped by lavas northwest of Crystal on the Defiance Plateau; Navajo="twin stars."

STAR LAKE, NM. Northeast of Crownpoint; Navajo=*Chéch'il Dah Łichíí*, meaning "oak red up."

STEAMBOAT CANYON, AZ. Navajo=*Hóyéé*, meaning "terrible," with reference to the eerie sound made by a spring in the canyon.

STONEY BUTTE, NM. North of Crownpoint; Navajo=*Tsé ł gaii*, meaning "white rock."

SUNSET CRATER, AZ. A recently eruptive volcanic cinder cone in a National Monument of that name northeast of Flagstaff, AZ; Navajo=*Dzi ł Bílá tah Łitsooí*, meaning "yellow top mountain."

SUNRISE, AZ. Settlement east of·Flagstaff, AZ; Navajo=*Séí Ndeesh-gizh*, meaning "sand gap."

SWEETWATER, AZ. Southeast of Mexican Water; Navajo=*Tó Łikan,* meaning "sweet water."

TAOS. Town in north-central New Mexico. The Spanish *taos* means "the badges of the orders of Saint Anthony and Saint John"; Navajo=*Tówoł,* meaning "gurgling water."

TAYLOR SPRING (west of Chambers, AZ). Navajo='*Asdzání Taah Yíyá,* meaning "the woman who went into the water."

TEEC NOS POS. The site of the original trading post on the Navajo Reservation in northeasternmost Arizona, known by its Navajo name, meaning "circle of cottonwood trees."

TES NEZ IAH TRADING POST. Trading center 39.5 miles east of Kayenta, AZ on U.S. Highway 160 in Navajo Country; name is Navajo for "tall cottonwood grove."

TESUQUE PUEBLO, NM. Navajo=*Tł'oh Łikizhí,* meaning "spotted grass."

THREE TURKEY HOUSE. Anasazi ruin near Canyon de Chelly; Navajo=*Chííł igai,* meaning "white ochre."

THOHEDLIH, NM. Opening of a box canyon a mile downstream from the confluence of the Los Piños and San Juan rivers, important locality in the Night Chant ceremonial; Navajo=*Tó 'Aheedlí,* meaning "water flows in a circle," probably an eddy in the river.

TIERRA AMARILLA, NM. Both the Spanish and the Navajo (*Łitsoí*) mean "yellow earth." A settlement south of Chama, NM, within the 600,000-acre Tierra Amarilla Land Grant from Mexico presented to Manuel Martinez in 1832, ownership having been handed down from Spain to Mexico and then the United States in 1848 by the Treaty of Guadalupe Hidalgo. Ownership was passed in 1909, after various land transactions, to the Arlington Land Company owned by John E. Andrus, Samuel S. Thorpe, and others. The land became known as "Alianza Country" (meaning alliance or coalition in Spanish) in the 1960s, when a political movement led by Reies Tijerina to free the land because of purported injustices to the local citizenry came to local violence.

TIJERAS CANYON. Spanish word for "scissors," and the canyon directly east of Albuquerque certainly appears to be a scissors-cut between the Sandia ("watermelon") Mountains to the north and the Manzano ("apple") Mountains to the south.

TOADLENA. Village at eastern edge of the Chuska Mountains; Navajo="flowing spring."

TOCITO, NM. South of Shiprock; Navajo="hot water."

TODILTO, NM. Navajo=*Tó Dildo'*, meaning "popping water." Name used for the Todilto Limestone or Formation (Jurassic).

TODILTO PARK. Circular valley north of Fort Defiance on the Defiance Plateau, an erosional expression of a volcanic vent (diatreme); Navajo="sounding water."

TOHACHI. Settlement and spring east of Chuska Peak; Navajo="water-is-scratched-for."

TOHONADLA. Spring south of the San Juan River near Bluff, UT; site of an oil field.

TOLANI LAKES, AZ. Navajo="many bodies of water."

TONALEA, AZ. Near Red Lake, northeast of Tuba City; Navajo="water flows and collects."

TONTO (Plateau): Spanish word for "fool" or "stupid."

TORREON. This village was given the Spanish name for "towers" built as fortifications to guard against attacks by Apache Indians.

TOWAOC. Headquarters for the Mountain Ute Indian Reservation south of Cortez, CO; Navajo=*Kin Dootł'izhí*, meaning "blue or green house."

TSAILE. Trading center on the Navajo Indian Reservation 20 miles east of Many Farms, south of Lukachukai, in northeastern Arizona; name is Navajo for "water-disappears-into-a-canyon" (Canyon del Muerto).

TSEGI (SEGI) TRADING POST. Trading post 11.4 miles southwest of Kayenta on U.S. Highway 160; name *(Tséyi')* is Navajo for "rock Canyon," and it is in a rocky canyon.

TUBA CITY. Founded by Mormons in 1872, who named it for Tueve, a Hopi leader from Oraibi whose name the Mormons mispronounced. The government bought them out in 1903 to establish Tuba City as an Indian Bureau headquarters. The Tuba Trading Post, built of native stone in 1905 in the circular shape of a Navajo hogan, is still in business today near the Navajo Reservation's only McDonalds, Taco Bell and Dairy Queen eating establishments. Also known as Tangled Water *(Tó Naneesdizí)* in Navajo, where the Water-is-close Clan originated.

TUNICHA MOUNTAIN. Near Lukachukai; Navajo=*Tó Ntsaa*, meaning "big water."

TURLEY, NM. Settlement east of Aztec, NM, on the south bank of the San Juan River; Navajo=*Náshdóí Bighan*, Meaning "wildcat's house."

TWIN LAKES, NM. Navajo=*Bahastł 'ah*, meaning "recess on each side of a hogan door."

TWO GRAY HILLS. Settlement in Chuska Valley, east of the Chuska Mountains, made famous for the high quality rugs of unique design made here. Navajo name is *Tsegilini*="stream-in-white-rocks" or *Bis Dah Łitso*, meaning "adobe-up-it is yellow"; also known as Crozier.

TWO WELLS, NM. South of Gallup; Navajo=*K'ai Nt'i'í*, meaning "willows extend in a line."

TYENDE MESA. Navajo=*Téé'ndééh*, referring to places where animals attempt to drink from depressions in the rock surface (natural tanks), and then can't get back out.

UTAH. Named for a tribe of Indians, the Utes, who dominated mountainous regions of the Southern Rocky Mountains when white settlers arrived. The name is from the Spanish *uta* given to the Native Americans by the early Spanish explorers, meaning "pimple-faced."

WALKER CREEK. A drainage off the northern Lukachukai Mountains near Sweetwater and Mexican Water, named for Captain Walker of Macomb's expedition; Navajo=*Tséyi' Hóchxo'i*, meaning "ugly canyon."

WALPI. A Hopi village on Black Mesa, AZ; Navajo=*Deez'áají'*, meaning "up to the point of the mesa."

WASHINGTON PASS (also Sonsela Pass). A much-travelled pass through the Chuska Mountains used by marauding Spaniards, Mexicans, and United States Army expeditions in the mid-1800s; named by Simpson in 1850 for Colonel Washington, a governor of New Mexico. Navajo=*Béésh Lich'íí'ii*, meaning "Cottonwood Pass." Recently renamed Narbona Pass by the Navajo Nation.

WATERFLOW, NM. Navajo=*Ch'ídíí ł chíí*, meaning "red devil."

WHEATFIELDS. Community on the west flank of the Chuska Mountains north of Fort Defiance and south of Lukachukai on the Defiance uplift; Navajo=*Tó Dzís'á*, meaning "a strip of water extending away into the distance"; called "Scattered Waters" in Navajo legends.

WHEATFIELDS CANYON, AZ. Near Wheatfields; Navajo=*Nát'ostse' 'Ál'íní*, meaning "place where stonepipes (for smoking) are made."

WHITE CONE, AZ. North of Indian Wells, AZ; Navajo=*Baa'oogeedí*, meaning "where it was dug into," or *Hak'eelt' izh*, meaning "head of a penis."

WHITEHORSE LAKE, NM. Navajo=*Tó Hwii ł híní*, meaning "where the water killed one."

WHITE HOUSE RUIN. Anasazi ruin in Canyon de Chelly National Monument; Navajo=*Kiníí' Na'ígai*, meaning "there is a white stripe across."

WHITE ROCK, AZ. Navajo=*Tsé 'A ł ch'í'*, Naagai meaning "white rocks descend together."

WINDOW ROCK, AZ. The Navajo capitol city northwest of Gallup, NM; Navajo=*Tségháhoodzání*, meaning "rock with a hole in it."

WILD HORSE MESA. North of Navajo Mountain, AZ; Navajo=*Tsé Ndoolzhah*, meaning "rock descending jagged."

WILLOW SPRING, AZ. Navajo=*'Aba'to'*, meaning "waiting water."

WINSLOW, AZ. Navajo=*Béésh Sinil*, meaning "metal objects lie," referring to the railroad.

WOMAN ROCK, AZ. Navajo=*Tsé 'Asdzáán*.

WUPATKI National Monument. Navajo=*'Anaasází Bikin*, meaning "alien ancestor's house."

YALE POINT. Northern headland of Black Mesa, south of Kayenta, AZ, named for Yale University.

YUCCA MOUNTAIN, AZ. Navajo=*Nooda Haas'áí*, meaning "yucca plants extend up out in a line," sometimes considered as a sacred mountain.

ZILDITLOI MOUNTAIN. A mountain in the Chuska Range, capped by columnar basalt; Navajo="mountain-with-hair-on-top."

ZUNI, NM. Town south of Gallup; Navajo=*Naasht'ézhí*, meaning "black streaked one."

ZUNI MOUNTAINS. A northwest-trending range of low mountains lying south of Grants and southeast of Gallup, NM. The range constitutes the erosional remnants of an elongate anticlinal structure (upfold) where metamorphic rocks of Precambrian age are exposed along the crest, overlain by Permian red beds. Thus, the fold is of Precambrian through Paleozoic age. Known as "Enemy Mountain" in Navajo mythology.

GLOSSARY

ANGULAR UNCONFORMITY. An unconformity or break between two series of rock layers such that rocks of the lower series underlie rocks of the upper series at an angle; the two series are not parallel. The lower series was deposited, then tilted and eroded prior to deposition of the upper layers.

ANTICLINE. An elongate fold in the rocks in which sides slope downward and away from the crest; an upfold.

ARKOSE. A sandstone containing a significant proportion of feldspar grains, usually signifying a source area composed of granite or gneiss.

BASEMENT. In geology, the crust of the Earth beneath sedimentary deposits, usually, but not necessarily, consisting of metamorphic and/or igneous rocks of Precambrian age.

BASEMENT FAULT. A fault that displaces basement rocks and originated prior to deposition of overlying sedimentary rocks. Such faults may or may not extend upward into overlying strata, depending upon their history of rejuvenation.

BASE LEVEL. The level, actual or potential, toward which erosion constantly works to lower the land. Sea level is the general base level, but there may be local, temporary base levels such as lakes.

BENTONITE. A rock composed of clay minerals and derived from the alteration of volcanic tuff or ash.

BRACHIOPOD. A type of shelled marine invertebrate now relatively rare but abundant in earlier periods of Earth history. They are common fossils in rocks of Paleozoic age. Brachiopods have a bivalve shell that is symmetrical right and left of center.

BRYOZOA. Tiny aquatic animals that build large colonial structures that are common as fossils in rocks of Paleozoic age.

CARBON-14 DATING OR RADIOCARBON DATING. A method of determining an age in years by measuring the concentration of carbon-14 remaining in formerly living matter, based on the assumption that assimilation of carbon-14 ceased abruptly at the time of death and that it thereafter remained a closed system. A half-life of 5570+/-30 years for carbon-14 makes the method useful in determining ages in the range of 500–40,000 years.

CEPHALOPOD. Marine mollusks that secrete shells that are chambered, usually coiled in a planospiral, but occasionally straight, with the main body of the animal housed in the last open chamber; they maintain buoyancy with gas filling the enclosed chambers and swim by jetting fluid; modern examples are the nautilus and squid. Ammonitic cephalopods have complexly crinkled chamber walls that may be used to distinguish species. They are especially useful in dating Mesozoic rocks.

CHERT. A very dense siliceous rock usually found as nodular or concretionary masses, or as distinct beds, associated with limestones. Jasper is red chert containing iron-oxide impurities.

CLASTIC ROCKS. Deposits consisting of fragments of preexisting rocks; conglomerate, sandstone, and shale are examples.

CONGLOMERATE. The consolidated equivalent of gravel. The constituent rock and mineral fragments may be of varied composition and range widely in size. The rock fragments are rounded and smoothed from transportation by water.

CONODONTS. Toothlike microfossils made of amber-colored calcium phosphate that may occur singularly or in mixed assemblages of variously shaped parts. They occur only in fine-grained rocks of marine origin and are believed to be derived from extinct annelid worms. As they evolved very rapidly, they are useful in dating and correlating rocks of Paleozoic and Mesozoic age.

CONTACT. The surface, often irregular, which constitutes the junction of two bodies of rock.

CONTINENTAL DEPOSITS. Deposits laid down on land or in bodies of water not connected with the ocean.

CORRELATION. The process of determining the position or time of occurrence of one geologic phenomenon in relation to others. Usually it means determining the equivalence of geologic formations in separated areas through a comparison and study of fossils or rock peculiarities.

CRINOID. Marine invertebrate animals, abundant as fossils in rocks of Paleozoic age. Most lived attached to the bottom by a jointed stalk, the "head" resembling a lily-like plant, hence the common name "sea lily."

DIATREME. A breccia-filled volcanic pipe that was formed by a gaseous explosion.

DIKE. A sheetlike body of igneous rock that filled a fissure in older rock while in a molten state. Dikes that intrude layered rocks cut the beds at an angle.

DISCONFORMITY. A break in the orderly sequence of stratified rocks above and below which the beds are parallel. The break is usually indicated by erosional channels, indicating a lapse of time or absence of part of the rock sequence.

DOLOMITE. A mineral composed of calcium and magnesium carbonate, or a rock composed chiefly of the mineral dolomite, formed by alteration of limestone.

DOME. An upfold in which strata dip downward in all directions from a central area; the opposite of a basin.

EOLIAN. Pertaining to wind. Designates rocks or soils whose constituents have been transported and deposited by wind. Windblown sand and dust (loess) deposits are termed eolian.

EROSIONAL UNCONFORMITY. A break in the continuity of deposition of a series of rocks caused by an episode of erosion.

EXTRUSIVE ROCK. A rock that has solidified from molten material poured or thrown out onto the Earth's surface by volcanic activity.

FACIES. Generally, this term refers to a physical aspect or characteristic of a sedimentary rock, as related to adjacent strata. It is usually applied to distinguish different aspects of the sediments in time-equivalent or laterally continuous beds. For example, the white sandstone facies of the Cedar Mesa Sandstone changes laterally to the age-equivalent red arkosic sandstone facies of the Cutler Group in Canyonlands Country. Such a change from one aspect to another is called a facies change.

FAULT. A break or fracture in rocks, along which there has been movement, one side relative to the other. Displacement along a fault may be vertical (normal or reverse fault) or lateral (strike-slip or "wrench" fault).

FORAMINIFERA. Generally microscopic one-celled animals (Protozoa), almost entirely of marine origin, with sufficiently durable shells capable of being preserved as fossils. They are usually abundant in marine sediments, and are sufficiently small to be retrievable in drill cuttings and cores.

FORMATION. The fundamental unit in the local classification of layered rocks, consisting of a bed or beds of similar or closely related rock types, and differing from strata above and below. A formation must be readily distinguishable, thick enough to be mappable, and of broad regional extent. A formation may be subdivided into two or more MEMBERS, and/or combined with other closely related formations to form a GROUP.

FUSULINIDS. Small spindle-shaped Foraminifera occurring as elongate chambers enrolled into complex internal forms that resemble a jellyroll. They are found only in marine rocks of Pennsylvanian and Permian age where they are excellent fossils for dating and correlating sedimentary rocks because of their rapid evolutionary history.

GEOLOGIC MAP. A map showing the geographic distribution of geologic formations and other geologic features, such as folds, faults, and mineral deposits, by means of color or other appropriate symbols.

GNEISS. A banded metamorphic rock with alternating layers of usually elongated tubular, unlike minerals.

GRANITE. An intrusive igneous rock with visibly granular, interlocking, crystalline quartz, feldspar, and perhaps other minerals.

IGNEOUS ROCK. Rocks formed by solidification of molten material (magma), including rocks crystallized from cooling magma at depth (intrusive), and those poured out onto the surface as lavas (extrusive).

INTRUSIVE ROCK. Rock that has solidified from molten material within the Earth's crust and has not reached the surface; it usually has a visibly crystalline texture.

LIMESTONE. A bedded sedimentary deposit consisting chiefly of calcium carbonate, usually formed from the calcified hard parts of organisms.

MASSIF. A massive topographic and structural uplift, commonly formed of rocks more rigid than those of its surroundings. These rocks are commonly protruding bodies of basement rocks, consolidated during earlier orogenies.

METAMORPHIC ROCK. Rocks formed by the alteration of preexisting igneous or sedimentary rocks, usually by intense heat and/or pressure, or mineralizing fluids.

MINETTE. A dark-colored intrusive igneous rock primarily composed of biotite phenocrysts (enlarged crystals) in a groundmass of orthoclase feldspar and biotite; it is commonly found in dikes associated with diatremes.

MORAINE. A mound, ridge, or other distinct accumulation of unsorted, unstratified drift, predominantly a heterogeneous mixture of mud, sand, gravel, and boulders, deposited by the melting of glacial ice.

OROGENY. Literally. the process of formation of mountains, but practically the processes by which structures in mountainous regions were formed, including folding, thrusting, and faulting in the outer layers of the crust, and plastic folding, metamorphism and plutonism (emplacement of magma) in the inner layers. An episode of structural deformation may be called an orogeny, e.g. the Laramide Orogeny.

SANDSTONE. A consolidated rock composed of sand grains cemented together; usually composed predominantly of quartz, it may contain other sand-size fragments of rocks and/or minerals.

SCHIST. A crystalline metamorphic rock with closely spaced foliation (platy texture) that splits into thin flakes or slabs.

SEDIMENTARY ROCK. Rocks composed of sediments, usually aggregated through processes of water, wind, glacial ice, or organisms, derived from preexisting rocks. In the case of limestones, constituent particles are usually derived from organic processes.

SHALE. Solidified mud, clays, and silts, that are fissile (split like paper) and break along original bedding planes.

SILL. A tabular body of igneous rock that was injected in the molten state concordantly between layers of preexisting rocks.

STRATIGRAPHY. The definition and interpretation of layered rocks, the conditions of their formation, their character, arrangements, sequence, age, distribution, and correlation, using fossils and other means.

STRATUM. A single layer of sedimentary rock, separated from adjacent strata by surfaces of erosion, non-deposition, or abrupt changes in character. Plural: strata.

SYNCLINE. An elongate, troughlike downfold in which the sides dip downward and inward toward the axis.

TECTONIC. Pertaining to rock structures formed by Earth movements, especially those that are widespread.

TRILOBITE. A general term for a group of extinct animals (arthropods) that occurs as fossils in rocks of Paleozoic age. The fossils consist of flattened, segmented shells with a distinct thoraxial lobe and paired appendages, usually found as partial fragments.

TYPE LOCALITY. The place from which the name of a geologic formation is taken and where the unique characteristics of the formation may be examined.

UNCONFORMITY. A surface of erosion or non-deposition separating sequences of layered rocks.

REFERENCES

General Interest

Agenbroad, Larry D. (1990) *Before the Anasazi: Early man on the Colorado Plateau.* Plateau, Museum of Northern Arizona.

Baars, Donald L. (1983) *The Colorado Plateau—A Geologic History.* University of New Mexico Press, Albuquerque.

————. (1989) *Canyonlands Country—Geology of Canyonlands and Arches National Parks.* Cañon Publishers Ltd. and Canyonlands Natural History Association, Lawrence, Kansas.

————. (1992) *The American Alps—The San Juan Mountains of Southwest Colorado.* University of New Mexico Press, Albuquerque.

Baars, Don, and Stevenson, Gene. (1986) *San Juan Canyons—A River Runner's Guide.* Cañon Publishers Ltd., Lawrence, Kansas.

Beus, Stanley S., and Morales, Michael. (1990) *Grand Canyon Geology.* Oxford University Press, New York and Museum of Northern Arizona Press, Flagstaff.

Cowie, J.W., and Bassett, M.G. (1989) *Global Stratigraphic Chart.* International Union of Geological Sciences, Bureau of International Commission on Stratigraphy (ICS:IUGS). Supplement to Episodes vol. 12, no. 2, Chart.

Christiansen, Paige W. (1989) *The Story of Oil in New Mexico.* New Mexico Bureau of Mines and Mineral Resources. Scenic Trips to the Geologic Past no. 14.

Evers, Larry, ed. (1982) *Between Sacred Mountains.* Sun Tracks and University of Arizona Press, Tucson.

Goodman, James M. (1982) *Navajo Atlas.* University of Oklahoma Press, Norman.

GRANT, CAMPBELL. (1978) *Canyon de Chelly—Its People and Rock Art.* University of Arizona Press, Tucson.

IVERSON, PETER. (1981) *The Navajo Nation.* University of New Mexico Press, Albuquerque.

MALLORY, WILLIAM W., ed. (1972) *Geologic Atlas of the Rocky Mountain Region.* Rocky Mountain Association of Geologists, Denver.

MCNITT, FRANK. (1990) *Navajo Wars.* University of New Mexico Press, Albuquerque.

NELSON, LISA. (1990) *Ice Age Mammals of the Colorado Plateau.* Northern Arizona University, Flagstaff. Report on studies by LARRY D. AGENBROAD and JIM I. MEAD.

OWEN, E.W. (1975) *Trek of the Oil Finders: A History of Exploration for Petroleum.* American Association of Petroleum Geologists Memoir 6.

UNDERHILL, RUTH. (1953) *Here Come the Navaho!* Department of the Interior, Bureau of Indian Affairs, Lawrence, KS.

ZOLBROD, PAUL G. (1984) *Diné Bahane'.* University of New Mexico Press, Albuquerque.

GEOLOGIC GUIDEBOOKS

ANDERSON, ORIN J., et al., eds. (1989) *Southeastern Colorado Plateau.* New Mexico Geological Society 40th Annual Field Conference Guidebook.

ANDERSON, ROGER Y., and HARSHBARGER, JOHN W., eds. (1958) *Black Mesa Basin, Northeastern Arizona.* New Mexico Geological Society 9th Field Conference Guidebook.

FASSETT, J.E., ed. (1977) *San Juan Basin III, Northwestern New Mexico.* New Mexico Geological Society 28th Field Conference Guidebook.

JAMES, H.L., ed. (1973) *Monument Valley and Vicinity, Arizona and Utah.* New Mexico Geological Society 24th Field Conference Guidebook.

LUCAS, SPENCER G., eds. (1992) *San Juan Basin IV.* New Mexico Geological Society 43rd Annual Field Conference Guidebook.

TRAUGER, FREDERICK D., ed. (1967) *Defiance-Zuni-Mt. Taylor Region, Arizona and New Mexico.* New Mexico Geological Society 18th Field Conference Guidebook.

Technical Publications

AKERS, J.P., COOLEY, M.E., and REPENNING, C.A. (1958) "Moenkopi and Chinle Formations of Black Mesa and adjacent areas." In R.Y. Anderson and J.W. Harshbarger, eds. *Guidebook of Black Mesa Basin, Northeastern Arizona.* New Mexico Geological Society 9th Field Conference Guidebook, 88–94.

ALLEN, J. E., and BALK, R. (1954) *Mineral Resources of Fort Defiance and Tohatchi Quadrangles, Arizona and New Mexico.* New Mexico Bureau of Mines and Mineral Resources Bulletin 36.

ARMSTRONG, A. K. (1967) *Biostratigraphy and Carbonate Facies of the Mississippian Arroyo Peñasco Formation, North-Central New Mexico.* New Mexico Bureau of Mines and Mineral Resources Memoir 20.

ARMSTRONG, A.K., and HOLCOMB, L.D. (1989) "Stratigraphy, facies and paleotectonic history of Mississippian rocks in the San Juan Basin of northwestern New Mexico and adjacent areas." In O.J. Anderson, et al., eds. *Southeastern Colorado Plateau.* New Mexico Geological Society Guidebook, 40th Field Conference Guidebook, 159–66.

ARMSTRONG, A.K., and MAMET, B.L. (1977) "Biostratigraphy and paleogeography of the Mississippian System in northern New Mexico and adjacent San Juan Mountains of southwestern Colorado." In J.E. Fassett and H.L. James, eds. *Guidebook of San Juan Basin III, Northwestern New Mexico.* New Mexico Geological Society 28th Field Conference Guidebook, 111–27.

BAARS, D.L. (1962) *Permian System of the Colorado Plateau.* American Association of Petroleum Geologists Bulletin vol. 46, 149–218.

————. (1966) Pre-Pennsylvanian paleotectonics—Key to basin evolution and petroleum occurrences in the Paradox basin. American Association of Petroleum Geologists Bulletin vol. 50, 2082–111.

————. (1973) "Permianland: The rocks of Monument Valley." In H.L. James, ed. *Guidebook of Monument Valley and vicinity, Arizona and Utah.* New Mexico Geological Society 24th Field Conference Guidebook, 68–71.

————. (1988) "Triassic and older stratigraphy; Southern Rocky Mountains and Colorado Plateau" In L.L. Sloss, ed. Sedimentary Cover—North American Craton, U.S.. Geological Society of America. The Geology of North America, vol. D-2, 53–64.

BAARS, D.L., et al. (1991) *Redefinition of the Pennsylvanian-Permian boundary in Kansas, Midcontinent, U.S.A.*. Program and Abstracts. International Congress on the Permian System of the World, Perm', USSR, A3.

―――. (1994) *Proposed repositioning of the Pennsylvanian-Permian boundary in Kansas.* Kansas Geological Survey Bulletin 231, 5–11.

BAARS, D.L., and STEVENSON, G.M. (1977) "Permian rocks of the San Juan Basin." In J.E. Fassett and H.L. James, eds. *Guidebook of San Juan Basin III, Northwestern New Mexico.* New Mexico Geological Society 28th Field Conference Guidebook, 133–38.

―――. (1981) "Tectonic evolution of the Paradox basin." In D.L. Weigand, ed. *Geology of the Paradox Basin.* Rocky Mountain Association of Geologists Guidebook, 23–31.

―――. (1982) "Subtle stratigraphic traps in Paleozoic rocks of the Paradox basin." In M. Halbouty, ed. *Deliberate Search for the Subtle Trap.* American Association of Petroleum Geologists Memoir 32, 131–58.

BAARS, D.L., and 15 others. (1988) "Basins of the Rocky Mountain region" In L.L. Sloss, ed. Sedimentary Cover—North American Craton, U.S.. Geological Society of America, The Geology of North America vol. D–2, 109–220.

BAKER, A.A., and REESIDE, J.B., JR. (1929) *Correlation of the Permian of Southern Utah, northern Arizona, northwestern New Mexico, and southwestern Colorado.* American Association of Petroleum Geologists Bulletin vol. 13, 1413–448.

BAKER, A.A., DANE, C.H., and MCKNIGHT, E.T. (1936) *Geology of the Monument Valley-Navajo Mountain region, San Juan county, Utah.* United States Geological Survey Bulletin 865, 106.

BEAUMONT, E.C., and READ, C.B. (1950) *Geologic History of the San Juan Basin area, New Mexico and Colorado.* New Mexico Geological Society 1st Field Conference Guidebook, 49–52.

BLAGBROUGH, J.W. (1967) "Cenozoic Geology of the Chuska Mountains." In F.D. Trauger, ed. *Guidebook of Defiance-Zuni-Mt. Taylor Region.* New Mexico Geological Society 18th Field Conference Guidebook, 70–77.

BLAKEY, R.C. (1990) *Stratigraphy and geologic history of Pennsylvanian and Permian rocks, Mogollon Rim region, central Arizona and vicinity.* Geological Society of America Bulletin vol. 102, 1189–1217.

————. (1993) "Supai Group and Hermit Formation." In S.S. Beus, and M. Morales, eds. *Grand Canyon Geology.* Oxford University Press and Museum of Northern Arizona Press, 147–82.

BREW, J.O., and HACK, J.T. (1939) "Prehistoric use of coal by Indians of northern Arizona." *Plateau,* vol. 12, 8–14.

CHENOWETH, W.L. (1967) "The uranium deposits of the Lukachukai Mountains, Arizona." In F.D. Trauger, ed. *Guidebook of Defiance-Zuni-Mt. Taylor Region.* New Mexico Geological Society 18th Field Conference Guidebook, 78–85.

————. (1977) "Uranium in the San Juan Basin-An overview." In J.E. Fassett and H.L. James, eds. *Guidebook of San Juan Basin III, Northwestern New Mexico.* New Mexico Geological Society 28th Field Conference Guidebook, 257–62.

————. (1989) "Ambrosia Lake, New Mexico—giant uranium district." In O.J. Anderson, et al., eds. *Southeastern Colorado Plateau.* New Mexico Geological Society 40th Field Conference Guidebook, 297–302.

CHENOWETH, W.L., and MALAN, R.C. (1973) "The uranium deposits of northeastern Arizona." In H.L. James, ed. *Guidebook to Monument Valley and Vicinity, Arizona and Utah.* New Mexico Geological Society 24th Field Conference Guidebook, 139–49.

CHUVASHOV, B.I. (1989) "The Carboniferous–Permian boundary in the USSR." In B.R. Wardlaw, ed. *Working Group on the Carboniferous-Permian boundary.* 28th International Geological Congress Proceedings, 42–56.

CONDON, S.M. (1989) "Modifications to Middle and Upper Jurassic nomenclature in the southeastern San Juan Basin, New Mexico" In O.J. Anderson, et al., eds. *Southeastern Colorado Plateau.* New Mexico Geological Society 40th Field Conference Guidebook, 231–38.

DAKE, C.L. (1920) "The Pre-Moenkopi unconformity of the Colorado Plateau." *Journal of Geology,* vol. 28: 61–74.

DAVYDOV, V.I., et al. (1991) *The Carboniferous-Permian boundary in the USSR and its correlation.* Program with Abstracts. International Congress on the Permian System of the World, Perm', USSR, A3.

DAVIS, G.H. (1978) "Monocline fold pattern of the Colorado Plateau." In V. Matthews, ed. *Laramide folding associated with basement block faulting in the western U.S..* Geological Society of America Memoir 151, 215–33.

DELANEY, PAUL T., and POLLARD, DAVID D. (1981) *Deformation of Host Rocks and flow of Magma during Growth of Minette Dikes and Breccia-Bearing Intrusions near Ship Rock, New Mexico.* United States Geological Survey Professional Paper 1202.

DUBIEL, R.F. (1989) "Sedimentology and revised nomenclature for the upper part of the Upper Triassic Chinle Formation and the Lower Jurassic Wingate Sandstone, northwestern New Mexico and northeastern Arizona." In O.J. Anderson, et al. eds. *Southern Colorado Plateau.* New Mexico Geological Society 4th Field Conference Guidebook, 213–23.

DUTTON, C.E. (1885) *Mount Taylor and the Zuni Plateau.* United States Geological Survey 6th Annual Report, Plate 16, 105–98.

ELLINGSON, J.A. (1973) "The Mule Ear diatreme." In D.L. Baars, ed. *Geology of the Canyons of the San Juan River.* Four Corners Geological Society Guidebook, Durango, Colorado.

ELSTON, W.E. (1960) *Structural development and Paleozoic stratigraphy of Black Mesa basin, northeastern Arizona and surrounding areas.* American Association of Petroleum Geologists Bulletin vol. 44, 21–36.

FASSETT, JAMES E., and HINDS, JIM S. (1971) *Geology and Fuel Resources of the Fruitland Formation and Kirtland Shale of the San Juan Basin, New Mexico and Colorado.* United States Geological Survey Professional Paper 676.

FITZSIMMONS, J.P. (1967) "Precambrian rocks of the Zuni Mountains." In F.D. Trauger, ed. *Guidebook of Defiance-Zuni-Mt. Taylor Region, Arizona and New Mexico.* New Mexico Geological Society 18th Field Conference Guidebook, 119–21.

———. (1973) "Tertiary igneous rocks of the Navajo Country, Arizona, New Mexico and Utah" In H.L. James, ed. *Guidebook to Monument Valley and Vicinity, Arizona and Utah.* New Mexico Geological Society 24th Field Conference Guidebook, 106–9.

GILBERT, G.K. (1875) *Report on the geology of portions of New Mexico and Arizona.* United States Geological and Geographical Survey West of the 100th Meridian vol. 3, 503–67.

GREEN, M.W. (1974) *The Iyanbito Member (a new stratigraphic unit) of the Jurassic Entrada Sandstone, Gallup-Grants area, New Mexico.* United States Geological Survey Bulletin 1395-D, D1–D12.

GREEN, M.W., and PIERSON, C.T. (1977) "A summary of the stratigraphy and depositional environments of Jurassic and related rocks in the San Juan Basin, Arizona, Colorado and New Mexico." In J.E. Fassett and H.L. James, eds. *Guidebook of San Juan Basin III, Northwestern New Mexico.* New Mexico Geological Society 28th Field Conference Guidebook, 147–52.

GREGORY, H.E. (1911) *The San Juan oil field, San Juan County, Utah.* United States Geological Survey Bulletin 431-A, 11–25.

———. (1917) *Geology of the Navajo Country.* United States Geological Survey Professional Paper 93.

———. (1938) *The San Juan Country.* United States Geological Survey Professional Paper 188.

HAGER, DORSEY. (1922) *Stratigraphy, northeast Arizona–southeast Utah.* Mining and Oil Bulletin vol. 8, no. 1, 26.

HAMBLIN, W.K. (1990) "Late Cenozoic lava dams in the western Grand Canyon" In S.S. Beus and M. Morales, eds. *Grand Canyon Geology.* Oxford University Press and Museum of Northern Arizona Press, 385–434.

HARSHBARGER, J.W., Repenning, C.A., and IRWIN, J.H. (1957) *Stratigraphy of the Uppermost Triassic and the Jurassic Rocks of the Navajo Country.* United States Geological Survey Professional Paper 291.

HOPKINS, R.L. (1990) "Kaibab Formation" In S.S. Beus and M. Morales, eds. *Grand Canyon Geology.* Oxford University Press and Museum of Northern Arizona Press, 225–46.

HUNT, A.P., and LUCAS, S.G. (1992) "Stratigraphy, paleontology and age of the Fruitland and Kirtland Formations, (Upper Cretaceous) San Juan Basin, New Mexico." In S.G. Lucas, et al. *San Juan Basin IV.* New Mexico Geological Society 43rd Field Conference Guidebook, 217–40.

IRWIN, J.H., STEVENS, P.R., and COOLEY, M.E. (1971) *Geology of the*

Paleozoic Rocks, Navajo and Hopi Indian Reservations, Arizona, New Mexico and Utah. United States Geological Survey Professional Paper 521–C.

JENTGEN, R.W. (1977) "Pennsylvanian rocks in the San Juan Basin, New Mexico and Colorado." In J.E. Fassett and H.L. James, eds. *Guidebook of San Juan Basin III, northwestern New Mexico.* New Mexico Geological Society 28th Field Conference Guidebook, 129–32.

KARNA, W.W. (1977) "Utah International's Navajo Mine." In J.E. Fassett, ed. *San Juan Basin III.* New Mexico Geological Society 28th Field Conference Guidebook, 251-52.

KELLEY, VINCENT C. (1955) *Regional Tectonics of the Colorado Plateau and Relationship to the Origin and Distribution of Uranium.* University of New Mexico Publications in Geology Number 5. University of New Mexico Press, Albuquerque.

————. (1958) "Tectonics of the Black Mesa Basin region of Arizona." In R.Y. Anderson and J.W. Harshbarger, eds. *Guidebook of the Black Mesa Basin, Northeastern Arizona.* New Mexico Geological Society 9th Field Conference Guidebook, 137–45.

————. (1967) "Tectonics of the Zuni-Defiance region, New Mexico and Arizona." In F.D. Trauger, ed. *Guidebook of Defiance-Zuni-Mt. Taylor Region, Arizona and New Mexico*: New Mexico Geological Society 18th Field Conference Guidebook, 28–31.

KITTEL, D.F., KELLEY, V.C. and MELANCON, P.E. (1967) "Uranium deposits of the Grants region." In F.D. Trauger, ed. *Guidebook of the Defiance-Zuni-Mt. Taylor Region*: New Mexico Geological Society 18th Field Conference Guidebook, 173–83.

KLINK, R.E. (1973) "Movie-making in Monument Valley." In H.L. James, ed. *Guidebook to Monument Valley and Vicinity, Arizona and Utah.* New Mexico Geological Society 24th Field Conference Guidebook, 199–203.

LANCE, J.F. (1958) "Precambrian rocks of northern Arizona." In R.Y. Anderson and J.W. Harshbarger, eds. *Guidebook of the Black Mesa Basin, Northeastern Arizona.* New Mexico Geological Society 9th Field Conference Guidebook, 66–70.

LIKHAREV, B.K. (1959) "The boundaries and principal subdivisions of the Permian System." *Soviet Geology.* vol. 6: 13–30.

Love, J.C., Hjellming, C.A., and Troxel, N.H. (1992) *Bibliography of New Mexico geology and mineral technology 1986*. New Mexico Bureau of Mines and Mineral Resources Bulletin 130.

Lucas, S.G. (1977) "Vertebrate paleontology of the San Jose Formation, east-central San Juan Basin, New Mexico." In J.E. Fassett, ed. *Guidebook of San Juan Basin III, northwestern New Mexico*. New Mexico Geological Society Guidebook, 221–25.

Lucas, S.G., and Hayden, S.N. (1989) "Triassic stratigraphy of west-central New Mexico." In O.J. Anderson, et al., eds. *Southeastern Colorado Plateau*: New Mexico Geological Society, 40th Field Conference, Guidebook 191–211.

Lucas, S.G., and Williamson, T.E. (1992) "Fossil mammals and the Early Eocene age of the San Jose Formation." In Lucas, et al., eds. *San Juan Basin IV*: New Mexico Geological Society 43rd Field Conference Guidebook, 311–16.

Lucchitta, Ivo. (1990) "History of the Grand Canyon and of the Colorado River in Arizona." In S.S. Beus and M. Morales, eds. *Grand Canyon Geology*. Oxford University Press, New York and Museum of Northern Arizona Press, Flagstaff, 311–32.

Mallory, W.W. (1972) "Pennsylvanian arkose and the Ancestral Rocky Mountains." In W.W. Mallory, ed. *Geologic Atlas of the Rocky Mountain Region*. Rocky Mountain Associaton of Geologists, Denver, Colorado, 131–32.

McKee, E.D. (1933) *The Coconino Sandstone—Its history and origin*: Carnegie Institute of Washington Publication no. 440, 77–115.

————. (1951) *Sedimentary basins of Arizona and adjacent areas*. Geological Society of America Bulletin vol. 62, 481–505.

————. (1954) *Stratigraphy and history of the Moenkopi Formation of Triassic age*. Geological Society of America Memoir 61.

————. (1958) "The Redwall Limestone" In R.Y. Anderson and J.W. Harshbarger, eds. *Guidebook of the Black Mesa Basin, Northeastern Arizona*. New Mexico Geological Society 9th Field Conference Guidebook, 74–77.

McLemore, V.T., and Chenoweth, W.L. (1992) "Uranium deposits in the eastern San Juan Basin, Cibola, Sandoval and Rio Arriba Counties, New Mexico." In S.G. Lucas, et al., eds. *San Juan Basin IV*: New Mexico Geological Society 43rd Field Conference Guidebook, 341–50.

MISER, H.D. (1924) *The San Juan Canyon, southeastern Utah.* United States Geological Survey Water-Supply Paper 538.

MOLENAAR, C.M. (1977) "Stratigraphy and depositional history of Upper Cretaceous rocks of the San Juan Basin area, New Mexico and Colorado, with a note on economic resources." In J.E. Fasset and H.L. James, eds. *Guidebook of San Juan Basin III, Northwestern New Mexico*: New Mexico Geological Society 28th Field Conference Guidebook, 159–66.

MOLENAAR, C.M. (1983). "Major depositional cycles and regional correlations of Upper Cretaceous rocks, southern Colorado Plateau and adjacent areas." In M.W. Reynolds and E.D. Dolly, eds. *Mesozoic paleogeography of west-central United States*. Rocky Mountain Section, Society of Economic Paleontologists and Mineralogists, 201–24.

MOLENAAR, C.M., and RICE, D.D. (1988) "Cretaceous rocks of the Western Interior Basin." In L.L. Sloss, ed. *Sedimentary Cover—North American Craton*. United States Geological Society of America. The Geology of North America vol. D-2, 77–82.

NOWELS, K.B. (1929) *Development and relation of oil accumulation to structure in the Shiprock District of the Navajo Indian Reservation, New Mexico*. American Association of Petroleum Geologists Bulletin vol. 13, 1119.

O'SULLIVAN, R.B. (1977) "Triassic rocks in the San Juan Basin of New Mexico and adjacent areas." In J.E. Fassett and H.L. James, eds. *Guidebook of San Juan Basin III, Northwestern New Mexico*. New Mexico Geological Society 28th Field Conference Guidebook, 139–146.

O'SULLIVAN, R.B., and GREEN, M.W. (1973) "Triassic rocks of northeast Arizona and adjacent areas." In H.L. James, ed. *Guidebook of Monument Valley and Vicinity, Arizona and Utah*. New Mexico Geological Society 24th Field Conference Guidebook, 72–78.

O'SULLIVAN, R.B., and CRAIG, L.C. (1973) "Jurassic rocks of northeast Arizona and adjacent areas." In H.L. James, ed. *Guidebook of Monument Valley and Vicinity, Arizona and Utah*. New Mexico Geological Society 24th Field Conference Guidebook, 79–85.

PAGE, H.G., and REPENNING, C.A. (1958) "Late Cretaceous stratigraphy of Black Mesa, Navajo and Hopi Indian Reservations, Arizona." In R.Y. Anderson and J.W Harshbarger, eds. *Guidebook of the Black Mesa Basin, Northeastern Arizona.* New Mexico Geological Society 9th Field Conference Guidebook, 115–22.

PEIRCE, H.W. (1958) "Permian sedimentary rocks of the Black Mesa Basin area." In R.Y. Anderson and J.W. Harshbarger, eds. *Guidebook of the Black Mesa Basin, Northeastern Arizona.* New Mexico Geological Society 9th Field Conference Guidebook, 82–87.

———. (1967) "Permian stratigraphy of the Defiance Plateau, Arizona." In F.D. Trauger, ed. *Guidebook of Defiance-Zuni-Mt. Taylor Region, Arizona and New Mexico*: New Mexico Geological Society 18th Field Conference Guidebook, 57–62.

PEIRCE, H. WESLEY, KEITH, STANTON B., and WILT, JAN CAROL. (1970) *Coal, Oil, Natural Gas, Helium, and Uranium in Arizona.* Arizona Bureau of Mines Bulletin 182.

PETERSON, FRED. (1988) "A synthesis of the Jurassic System in the southern Rocky Mountain region." In L.L. Sloss, ed. *Sedimentary Cover—North American Craton; U.S..* Geological Society of America, The Geology of North America, vol. D-2, 65–76.

PETERSON, FRED, and KIRK, A.R. (1977) "Correlation of Cretaceous rocks in the San Juan, Kaiparowits and Henry Basins, southern Colorado Plateau." In J.E. Fassett and H.L. James, eds. *Guidebook of San Juan Basin III, Northwestern New Mexico.* New Mexico Geological Society 28th Field Conference Guidebook, 167–78.

PETERSON, FRED, and PIPIRINGOS, G.N. (1979) *Stratigraphic Relationships of the Navajo Sandstone to Middle Jurassic Formations, Southern Utah and Northern Arizona.* United States Geological Survey Professional Paper 1035-B.

POHLMANN, H.F. (1967) "The Navajo Indian Nation and Dineh Bi Keyah." In F.D. Trauger, ed. *Guidebook of Defiance-Zuni-Mt. Taylor Region, Arizona and New Mexico.* New Mexico Geological Society 18th Field Conference Guidebook, 63–69.

POTOCHNIK, A.R., and REYNOLDS, S.J. (1990) "Side canyons of the Colorado River, Grand Canyon." In S.S. Beus, and M. Morales, eds. *Grand Canyon Geology.* Oxford University Press and Museum of Northern Arizona Press, 461–81.

POWELL, J.W. (1875) *Exploration of the Colorado River of the West and its tributaries.* U.S. Government Printing Office, Washington, D.C., 3–145.

READ, C.B., and WANEK, A.A. (1961) *Stratigraphy of Outcropping Permian Rocks in Parts of Northeastern Arizona and Adjacent Areas.* United States Geological Survey Professional Paper 374-H.

REPENNING, C.A., LANCE, J.F., and IRWIN, J.H. (1958) "Tertiary stratigraphy of the Navajo Country." In R.Y. Anderson and J.W. Harshbarger, eds. *Guidebook of the Black Mesa Basin, Northeastern Arizona.* New Mexico Geological Society 9th Field Conference Guidebook, 123–29.

RIGBY, J. KEITH, JR. (1982) "Camarasaurus cf. supremus from the Morrison Formation near San Ysidro, New Mexico—The San Ysidro Dinosaur." In J.A. Grambling and S.G. Wells, eds. *Albuquerque Country II.* New Mexico Geological Society Guidebook, 271–72.

SAUCIER, A.E. (1967) "The Morrison Formation in the Gallup region." In F.D. Trauger, ed. *Guidebook of the Defiance-Zuni-Mt. Taylor Region.*: New Mexico Geological Society 18th Field Conference Guidebook, 138–44.

SEARS, JULIAN D. (1956) *Geology of Comb Ridge and Vicinity North of San Juan River, San Juan County, Utah.* United States Geological Survey Bulletin 1021-E.

SMITH, M.C., JR. (1967) "The AEC and the Grants Mineral Belt." In F.D. Trauger, ed. *Guidebook of the Defiance-Zuni-Mt. Taylor Region.* New Mexico Geological Society 18th Field Conference Guidebook, 184–87.

SMITH, C.T. (1967) "Jurassic stratigraphy of the north flank of the Zuni Mountains." In F.D. Trauger, ed. *Guidebook of Defiance-Zuni-Mt. Taylor Region:* New Mexico Geological Society 18th Field Conference Guidebook, 132–37.

STEVENSON, G.M., and BAARS, D.L. (1977) "Pre-Carboniferous paleotectonics of the San Juan Basin." In W.E. Fassett and H.L. James, eds. *San Juan Basin III.* New Mexico Geological Society 28th Field Conference Guidebook, 99–110.

———. (1986) "The Paradox: A pull-apart basin of Pennsylvanian age." In J.A. Peterson, ed. *Paleotectonics and Sedimentation.* American Association of Petroleum Geologists Memoir 41, 513–39.

STEWART, J.H. (1959) "Stratigraphic relations of Hoskinnini Member (Triassic?) of Moenkopi Formation on the Colorado Plateau." *American Association of Petroleum Geologists Bulletin* vol. 43, 1852–68.

STOKES, W.L. (1951) *Carnotite Deposits in the Carrizo Mountains Area, Navajo Indian Reservation, Apache County, Arizona, and San Juan County, New Mexico.* United States Geological Survey Circular 111.

———. (1973) "Geomorphology of the Navajo Country." In H.L. James, ed. *Guidebook of Monument Valley and Vicinity, Arizona and Utah.* New Mexico Geological Society 24th Field Conference Guidebook, 61–67.

VAUGHN, P.P. (1973) "Vertebrates from the Cutler Group of Monument Valley and vicinity." In H.L. James, ed. *Guidebook of Monument Valley and Vicinity, Arizona and Utah.* New Mexico Geological Society 24th Field Conference Guidebook, 99–105.

WATSON, E.L. (1973) "Navajo history: A 3000-year sketch." In H.L. James, ed. *Guidebook to Monument Valley and Vicinity, Arizona and Utah.* New Mexico Geological Society 24th Field Conference Guidebook, 181–85.

WENGERD, S.A. (1955) *Biohermal trends in Pennsylvanian strata of San Juan Canyon, Utah.* Four Corners Geological Society, Guidebook, First Field Conference, Geology of parts of the Paradox, Black Mesa, and San Juan Basins, 70–77.

———. (1955) *Geology of the Mexican Hat Oil Field, San Juan County, Utah.* Four Corners Geological Society Guidebook, First Field Conference, Geology of parts of the Paradox, Black Mesa, and San Juan Basins, 150–63.

———. (1962) "Pennsylvanian sedimentation in Paradox Basin, Four Corners Region" In Carl C. Branson, ed. *Pennsylvanian System in the United States.* American Association of Petroleum Geologists. A symposium, 264–330.

———. (1973) "Regional stratigraphic control of the search for Pennsylvanian petroleum, southern Monument Upwarp, southeastern Utah." In H.L. James, ed. *Guidebook to Monument Valley and Vicinity, Arizona and Utah.* New Mexico Geological Society 24th Field Conference Guidebook, 122–38.

WENGERD, S.A., and STRICKLAND, J.W. (1954) *Pennsylvanian stratig-*

raphy of the Paradox Basin, Four Corners region, Colorado and Utah.: American Association of Petroleum Geologists Bulletin vol. 38, 2157–99.

WENGERD, S.A., and MATHENY, M.L. (1958) *Pennsylvanian System of the Four Corners region.* American Association of Petroleum Geologists Bulletin vol. 42, 2048–106.

WILLIAMSON, T.E., and LUCAS, S.G. (1992) "Stratigraphy and mammalian biostratigraphy of the Paleocene Nacimiento Formation, southern San Juan Basin, New Mexico." In S.G. Lucas, et al., eds. *San Juan Basin IV.* New Mexico Geological Society 43rd Field Conference Guidebook, 265–96.

WOODWARD, L.A. (1973) "Structural framework and tectonic evolution of the Four Corners Region of the Colorado Plateau." In H.L. James, ed. *Guidebook of Monument Valley and Vicinity, Arizona and Utah.* New Mexico Geological Society 24th Field Conference Guidebook, 94–98.

WOODWARD, L.A., and CALLENDER, J.F. (1977) "Tectonic framework of the San Juan Basin." In J.E. Fassett and H.L. James, eds. *Guidebook of San Juan Basin III, Northwestern New Mexico.* New Mexico Geological Society 28th Field Conference Guidebook, 209–12.

YOUNG, R.G. (1973) "Cretaceous stratigraphy of the Four Corners Area." In H.L. James, ed. *Guidebook of Monument Valley and Vicinity, Arizona and Utah.* New Mexico Geological Society 24th Field Conference Guidebook, 86–93.

YOUNG, ROBERT W. (1961) *The origin and development of Navajo Tribal government. The Navajo Yearbook.* Window Rock, AZ, vol. 8, 371–11.

INDEX

San Juan Mountains, 11, 18, 20, 22, 25, 27, 29–31, 64, 107, 117, 119, 183, 184, 186
San Juan Oil Field, 153, 154
San Juan River, 5, 6, 3, 36, 40, 55, 66, 69, 70, 80, 82, 107, 108, 111, 112, 117, 118, 121, 123, 127, 152–54, 161, 163, 172, 184
San Juan Trading Post, 112
San Luis Valley, 8
San Pedro Mountain, 36
San Pedro Mountains, 3, 12
San Rafael Group, 62, 66, 107, 119
San Rafael Swell, 3, 34, 47, 62, 65, 90
San Ysidro, 44, 64, 66, 88, 179
San Ysidro Member, 44, 88
Sand Island, 107, 118, 119
Sand Island Recreational Area, 118
Sanostee, 179, 181
Santa Fe, 6, 25, 155, 157, 158, 165
Santa Fe Corporation, 157
Santa Fe Energy Company, 6
Santa Fe Railroad, 6, 165
satin spar, 18
Sawatch Range, 34
Sawmill, 91, 95, 112
schist, 21–24, 146
Schnebly Hill Formation, 43
Schrader, 152
Sedgwick, 15, 16
sedimentary rock, 5, 6, 15, 18, 19, 24, 129, 144–46
sedimentary rocks, 6, 10, 15, 17–19, 21, 23, 24, 27–29, 34, 39, 62, 68, 70, 71, 75, 91, 105, 123, 145, 149, 163, 164, 176–78
Segisaurus, 62
selenite, 18
Semionotis, 62
Seven Lakes, 153, 155
Shaler, 152
shark teeth, 73
Sheep Springs, 95, 99
Sheer Wall Rapid, 140
Shell Oil Company, 6, 30, 152, 161–63
Shinarump, 49–52, 89, 93, 100, 101, 113, 114, 159, 177, 180

Shinumo Quartzite, 24, 145
Ship Rock, 82, 171, 187
Shiprock, 6, 11, 74, 79, 81, 82, 99, 114, 153, 155–57, 172, 178, 181
Shivwits, 132
Siberia, 186
Sierra Nevada, 69
silica, 52, 53, 122
silicates, 37
Silurian, 15, 28, 141, 15
Simpson, 99
sinkholes, 32
Sis Naajini, 5, 7
Slave Province, 22
Sleeping Ute, 6
Slickhorn Gulch, 125, 127, 154
Slickhorn Rapid, 127
sloths, 185
Smith, 66
snails, 62
Snowdon Peak, 30
Soap Creek Rapid, 139
Sockdolager Rapid, 146
Soda Basin, 123
Sonsela Buttes, 99
Southern Union Gas Company, 160
Southwest Oil Company, 159
Spaniards, 185
Spider Rock, 87, 101, 103, 187
Spider Woman, 103, 187
spores, 38, 62, 89
Springdale Sandstone Member, 58
Stanolind, 162
Stevenson, 102
stratigraphy, 20, 39, 44, 64, 78, 153
Summerville Formation, 65, 90
Sumner, 61, 94, 186
Sun, 7, 21, 52, 79, 162, 187
Supai Formation, 25, 42, 87, 88, 94, 95, 101
Supai Group, 36, 41, 88, 140
Superior, 162
Surprise Valley Formation, 32
syenite, 163, 164
sylvite, 18
syncline, 40, 105, 110–12, 123, 125, 149, 154